Biology as a Social Weapon

The Ann Arbor Science for the People Editorial Collective

Ann Arbor, Michigan

Burgess Publishing Company
Minneapolis, Minnesota

Printed in the United States of America
Library of Congress Catalog Card Number 77-70866
ISBN: 0-8087-4534-4

0 9 8 7 6 5 4 3 2

Preface

For the most part, the papers presented in this book are the result of a symposium entitled "Biological Determinism: A Critical Appraisal," which was held at the University of Michigan on September 29th through October 3rd, 1975. The papers "The XYY Male: The Making of a Myth" and "Sociobiology: A New Biological Determinism" were added to the original papers for this publication.

The program was organized by the Ann Arbor chapter of Science for the People. Science for the People (also known as Scientists and Engineers for Political Action [SESPA]) is a national organization of scientists, engineers, and technical workers dedicated to progressive social change. The organization publishes a monthly magazine called *Science for the People* in which the issues raised in this book, as well as many other social and political issues in science and technology, are aired. The national office is located in the Boston area (897 Main Street, Cambridge, Mass. 02139).

The original symposium was assisted financially by the University of Michigan Values Program, the department of human genetics, the University of Michigan Medical School, and International Woman's Year.

The production of *Biology as a Social Weapon* was undertaken by the Editorial Collective of the Ann Arbor chapter of Scienҫe for the People (Douglas Boucher, Kathryn Dewey, Robert Noonan, Steven Risch, Scott Schneider, Jean Stout, and John Vandermeer). The introduction to the book, as well as the introductions to the individual topics and the epilogue, were prepared by the Editorial Collective.

Contents

Introduction

How often do we hear phrases like "blood is thicker than water," "that's just human nature," "it's woman's intuition," "they sure have got rhythm," "a woman's natural place is in the home," "he comes from good stock," and "those people breed like flies"? Do all these clichés have something in common?

How often do we feel that strong emotions, such as jealousy and selfishness, are deeply ingrained parts of our "human nature"? Have you ever felt that you were just born stupid (or smart)? Do such emotions have a common theme?

Can one help but feel that these cliches, and the emotions they provoke, stem from something very fundamental? Something biological? Something that is at the foundation of human nature?

There is an old and continuing theory that ties such observations together, a theory that "explains" selfishness, stupidity, brilliance, jealousy, love, and so on, in terms of biological imperatives. This theory has recently been called *biological determinism* and, although it is not new, it is receiving renewed attention because of supposedly new scientific analysis. Biological determinism attempts to explain much of the present human condition as a natural outgrowth of biological phenomena. Human nature is explained solely in terms of Darwinian evolution, gene expression, hormonal reality, or other purely biological processes. The contributors to this volume argue that such explanations have no scientific basis. This is a difficult task because so many people have internalized these explanations, whether consciously or unconsciously, and strongly hold them as basic tenets.

In order for you to gain a feeling for how difficult it is to dislodge such deeply felt opinions, we ask you to imagine yourself living in the fifteenth century. During the fifteenth century, all rational people firmly believed that the sun revolved around the earth. When you walk outside in the morning, you see the sun rise in the east, and you watch it gradually move across the sky until it sets in the west in the evening. Clearly, there can be no question that the sun revolves around the earth. Generations of people have watched the sun revolve around the earth, and you would no more consider questioning that fact than the fact that water runs downhill. Can you imagine how you might feel were someone to suggest that, in fact, the sun does not revolve around the earth? Your response might be similar to your response to someone telling you that water runs uphill, not downhill.

Our deeply ingrained feelings about human nature are analogous to the opinions that people in the fifteenth century formed when they observed the sun moving across the sky. Such feelings and opinions are based on observations that are difficult to deny. Although the theories of biological determinism are largely incorrect, their apparent ability to explain commonly observed human situations makes it difficult to demonstrate convincingly their incorrectness. We expect the analyses in this volume to generate strong reaction. Many will find the arguments unpalatable, for the arguments challenge some of our basic beliefs. We ask only that the reader consider the discussions in this book with an open mind, that the arguments be given as fair a consideration as possible, and that the subject matter itself be taken very seriously. For the matters we are concerned with are crucially important in today's world. Their implications are more far-reaching than most readers will realize initially.

Consider, for example, how readily the philosophy of Jensenism was taken up by makers of important government policy. Arthur Jensen's famous paper, "How Much Can We Boost IQ and Scholastic Achievement?"(1), attempted to show "scientifically" that the difference in IQ scores between blacks and whites is genetically determined. That paper did not remain solely in scholarly journals. It was read in its entirety into the *Congressional Record* by Congressman John R. Rarick(2). Daniel Moynihan, then special assistant to President Richard Nixon, said in 1970 that "the winds of Jensen" were gusting through the capital at gale force(3). John Neary, in a June 12, 1970, *Life* magazine article(3; p. 58B), quoted a "high government official" as telling him that Jensen's paper had "kicked up a lot more private reaction than you'd think. It's not something that anybody does talk about. It's *secret knowledge* in Washington, something everybody knows and doesn't say. In the bureaucracy, when they see these compensatory programs not working, they just look at each other." [Italics in original.] The notion that some people are stupid and others are bright due to their genetic makeup clearly has grave consequences for our vision of a just society.

The issue of IQ genetics is but one of the numerous aspects of biological determinism. This volume includes critical analyses of many of these deterministic theories—their history, social setting, evidence, conclusions, and ethical implications. In very general terms, these critical analyses proceed along two interrelated lines: first, why deterministic theories are scientifically unfounded, and, second, why the theories remain popular despite their lack of scientific basis. Metaphorically speaking, the critiques show that, first, contrary to popular opinion, the earth revolves around the sun and not vice versa and, second, there are reasons, usually political, why we are told that the sun revolves around the earth.

The book begins with an historical overview. In "Biological Determinism as a Social Weapon," Richard Lewontin notes that, although Western industrialized societies supposedly are based upon the principle of equal opportunity for all, it is clear that the distribution of power and wealth is grossly inequitable. The contradiction between the espoused goal and the apparent result has necessitated an explanation. Historically, the ideology of biological determinism has provided one by "proving" scientifically that those who receive a disproportionately small fraction of society's products do so because they are inherently inferior. Hence, we see arguments that blacks are poor because they are genetically less intelligent and that women generally do not hold responsible jobs because of their hormones. History is replete with examples of famous scientists making statements about human behavior that were demonstrably false at the time they were made. These assertions usually serve to support an unjust social ideology. Biological deterministic arguments are based on two unproven assumptions:

first, that there are complex social behaviors common to all people and that these behaviors collectively constitute a biologically determined "human nature" and, second, that differences in social behavior among individuals or groups are rooted in their biology. Neither assumption is supported by any scientific evidence.

Arthur Schwartz, in "The Politics of Statistics: Heredity and IQ," first discusses the nature of IQ tests and the concept of intelligence. He describes the ambiguity and cultural bias of these tests, showing how the politics of the testing movement have influenced their composition. Challenging many of the assumptions that go into the original collection of data and pointing out the blatant misinterpretation of much of these data, he concludes that there exists no evidence to support the notion that intelligence has a hereditary component.

Val Woodward follows by noting that the use of racial differences in IQ scores to justify racism "scientifically" has a long tradition, despite the clear fallacies in the argument. He reviews the use of IQ tests in the United States and points out how scientists' support of racist conclusions based on these tests led to the restriction of emigration from eastern and southern Europe and to the sterilization laws in many states. More recently, the argument has been revived, with no more scientific basis than before, by Arthur Jensen, William Shockley, and others. Woodward shows that all of Jensen's major points are false. He argues that a movement to combat scientific racism must necessarily be one for radical social change.

In the mid-sixties, it was claimed that males who had an extra Y chromosome (XYY males) were more likely to be aggressive and to become criminals. Many studies were done among prison populations to prove this hypothesis. Pyeritz et al., in their paper "The XYY Male: The Making of a Myth," take a critical look at how those studies were done and at the conclusions that were reached. The authors scrutinize the use of the research to perpetuate and legitimize unfounded ideas and they show how those ideas have profound political and social implications. As they state, "If crime and aggression are indeed biological, then social factors such as poverty and oppression, which are often associated with crime, can be ignored in favor of a medical solution."

Richard Kunnes, in "Political Determinants of Violence," points out that the biological deterministic concept that aggression is fundamentally a biological phenomenon takes attention away from those *social* forces that cause violence to individuals in many ways. The deterministic argument concerns itself solely with individual antisocial acts, obfuscating the more significant social violence of war, poverty, disease, and pollution. Such an ideology attempts to sell the notion that violent people are somehow genetically different. Thus, it helps to prevent people from taking action against social violence and aids those few who stand to gain from allowing, and even encouraging, social violence.

In "Science and Sex Roles in the Victorian Era," Robin Jacoby analyzes the Victorian period, an appropriate era for examining the connections between cultural norms of sex roles and scientific thinking. At that time, the dominant interest in the scientific method led scientists and doctors to develop justifications for the "natural" role of women, based on supposed biological facts. Women were claimed to be physically and intellectually inferior because of their smaller body and brain size and the "debilitating" effects of menstruation. Repressive attitudes towards sexuality reinforced these views about sex roles. The theories propounded by Victorian scientists, such as the idea that the intellectual development of a woman stunts her reproductive system, sound absurd to us today. However, they illustrate dramatically the way in which scientists move unhesitatingly from scientific data to interpretations of those data that are reflections of social ideology rather than the result of pure, unbiased scientific reasoning.

Pauline Bart discusses the question of gender behavior and its origins in "Biological Determinism and Sexism: Is It All in the Ovaries?" She notes that we are entering a conservative era in which biological explanations are used to account for differences in sex roles. Such explanations are challenged by much recent research. Nevertheless, they are still prevalent in gynecological textbooks and throughout medical practice today. A discussion of sex roles in the Israeli kibbutz is presented to illustrate social and economic reasons why women reverted to traditional female behavior in a supposedly egalitarian community.

John Vandermeer writes of the tendency to use current ecological theory to place the blame for environmental problems on "overpopulation" rather than on corporations and government. He calls such a tendency "ecological determinism." He particularly attacks the new theories of "triage" and the "lifeboat ethic" of Garrett Hardin, likening them to the ideology of Nazi Germany. Finding inadequate the more liberal interpretations of the social causes of so-called overpopulation, Vandermeer argues that the issue of overpopulation was developed by the powerful to distract people from the real problem of the inequitable distribution of resources. Finally, he notes a recent tendency among multinational corporations to utilize the concept of complexity of ecosystems as an excuse to gain the position of benevolent world manager. Such a development would only exacerbate an already critical problem.

Our society's view of nature as a natural resource stems from a social system in which people, like nature, are seen as objects to be used, according to Murray Bookchin in "Ecology, Society, and the Myth of Biological Determinism." All societies, in fact, tend to project their own social system onto nature, and in return this view of nature is used to justify the social system as "natural." For example, our market society has tended to see nature as a jungle in which ruthless competition prevails. A truly ecological society, as opposed to our present environmentalism, can only come about from a harmonious relationship among people.

The Boston Science for the People Sociobiology Study Group contributes a detailed critique of the most recent work of biological determinism, E. O. Wilson's *Sociobiology*. Wilson's descriptions of universals in human behavior, which not coincidentally are quite similar to the behavior of persons in present-day upper-class society, are not substantiated by a critical examination of the ethnographic literature. However, application of modern evolutionary theories, including kin selection and reciprocal altruism, allows one to explain essentially any human behavior as a biologically evolved adaptation. This method, of course, must ignore the total lack of evidence for genes controlling complex human behaviors. Furthermore, it is necessary to postulate *ad hoc* processes to explain contradictions between the observed rapidity of cultural evolution, with the resulting divergent cultures, and our knowledge of the process of natural selection from modern population genetics. The result is a theory that explains the *status quo* as "natural" and constricts the possibilities of change.

The issue of biological determinism shares elements with the old religious issue of freedom of the will; it is a doctrine of grace, as Lewontin points out, or a new version of the doctrine of Original Sin. Biological determinism puts limits on human freedom and is, therefore, necessarily a political issue. It says that, by their very nature, blacks cannot be scientists, women cannot be presidents, nations cannot be peaceful, and societies cannot live in harmony with nature. It tells us that it is not natural for people to live together in justice and friendship.

The contributors to this volume argue that human potentialities are boundless. They say that the question is not what we can do, but what we will do. Their message is that our possibilities are limited not by our biological makeup but by the social, economic, and political institutions of society, of which science is one of the most important. They tell us, contrary to the ideology of biological determinism, that humanity can be free.

References

1. Jensen, A. R. 1969. How much can we boost IQ and scholastic achievement? *Harvard Educational Review* 39:1-123.
2. U.S. Congress, House, *Congressional Record,* 91st Congress, 1st session, May 28, 1969, 115, part 2:14189-217.
3. Neary, J. 1970. Jensenism: Variations on a racial theme. *Life* 68(22):58B-65.

Biological Determinism as a Social Weapon

Richard C. Lewontin

The struggle between those who possess social power and those who do not, between "freeman and slave, patrician and plebian, lord and serf, guildmaster and journeyman, in a word, oppressor and oppressed"(1) is a war fought with many and varied weapons. Of highest importance are ideas, weapons in an ideological warfare by which every class struggling to maintain its grip on the world tries to justify its position morally and rationally, while those fighting to overturn the social order produce their own self-justificatory ideology as a counter weapon. If the revolution succeeds, that revolutionary ideology becomes transformed into a weapon of consolidation and conservation whereby yet further revolutionary challenges to the new dominant class can be resisted. Nothing better illustrates the historical progression of such ideological weapons than the revolution that created the twentieth-century market-industrial society.

The society of Europe before the seventeenth century (with the exception of certain mercantile Italian republics) was characterized by a static, aristocratic scheme of relations in which both peasants and landowners were bound to each other and to the land and in which changes in the social positions of individuals were exceedingly rare. Persons were said to owe their position in the world to the grace of God or to the grace of earthly lords. Even kings ruled *Deo gratia,* and changes in position could only occur by exceptional conferrals or withdrawals of divine or royal grace. But this rigid hierarchy directly obstructed the expansion of both mercantile and manufacturing interests who required access to political and economic power based on their entrepreneurial activities rather than on noble birth.

Moreover, the inalienability of land and the traditional guarantee of access to common land inhibited the rapid expansion of primary production and also maintained a scarcity of labor for manufactories. In Britain, the Acts of Enclosure of the eighteenth century broke this rigid system by allowing landlords to enclose land for wool production and simultaneously displacing tenants, who then became the landless industrial work force of the cities. At the same time in France, the old "nobility of the sword" was being challenged by the administrative and legal hierarchy who became the "nobility of the robe" and by the rich commoners of banking and finance. The bourgeois revolution was brewing, a revolution that was to break assunder the static feudal-aristocratic bonds and create instead an entrepreneurial society in which labor and money could more freely adapt to the demands of a rising commercial and industrial middle class. But the bourgeois revolution required an ideology justifying the assault on the old order and providing the moral and intellectual underpinnings of the new. This was the ideology of freedom, of individuality, of works as opposed to grace, and of equality and the inalienable rights to "life, liberty, and the pursuit of happiness." Paine, Jefferson, Diderot, and the Encyclopedists were the ideologues of the revolution, and one theme comes through in their writings: the old order was characterized by artificial hierarchies and artificial barriers to human desire and ambitions and those artificial barriers must be destroyed so that each person can take his or her natural place in society, according to his or her desire and ability. This is the origin of the idea of the "equal opportunity society" in which we are now supposed to live.

Yet, the bourgeois revolution that destroyed those artificial barriers seems not to have dispensed with inequality of station. There are still rich and poor, powerful and weak, both within and between nations. How is this to be explained? We might suppose that the inequalities are structural, that the society created by the revolution has inequality built into it and even depends upon that inequality for its operation. But that supposition, if taken seriously, would engender yet another revolution. The alternative is to claim that inequalities reside in properties of individuals rather than in the structure of social relations. This is the claim that our society has produced about as much equality as is humanly possible and that the remaining differences in status and wealth and power are the inevitable manifestations of *natural* inequalities in individual abilities. It is this latter claim that has been incorporated from an early stage into the ideology of the bourgeois revolution and that remains the dominant ideology of market-industrial societies today. Such a view does not threaten the *status quo* but, on the contrary, supports it by telling those who are without power that their position is the inevitable outcome of their own innate deficiencies and that, therefore, *nothing can be done about it.* A remarkably explicit recent statement of this assertion is that of Richard Herrnstein, a psychologist who is one of the leading ideologues of "natural inequality": "The privileged classes of the past were probably not much superior biologically to the downtrodden, which is why revolution had a fair chance of success. By removing artificial barriers between classes, society has encouraged the creation of biological barriers. When people can take their natural level in society, the upper classes will, by definition, have greater capacity than the lower"(2; p.221).

Here the entire scheme is laid out. The bourgeois revolution succeeded because it was only breaking down artificial barriers, but the remaining inequalities cannot be removed by a further revolution because what is left is the residue of biological differences that are ineradicable. We are not told precisely what principle of biology guarantees that biologically "inferior" groups cannot seize power from biologically "superior" groups, but the conceptual and factual errors of such a statement are irrelevant to its function. It is meant to convince us that, although we may not live in the best of all *conceivable* worlds, we live in the best of all *possible* worlds.

An important corollary, developed in nineteenth-century sociology, was that the natural sorting process that takes place in a free society is greatly aided by education since education is the means of bringing into actuality the latent differences among individuals. Lester F. Ward, the giant of nineteenth-century American sociology, wrote: "Universal education is the power which is destined to overthrow every species of hierarchy. It is destined to remove all artificial inequality and leave the natural inequalities to find their true level. The true value of a newborn infant lies . . . in its naked capacity for acquiring the ability to do"(3). (It is the same L. F. Ward who in his *Pure Sociology*(4) claimed that it was more permissible for a man of a superior race to rape a woman of an inferior race than *vice versa* because it would be a leveling up rather than a leveling down!)

Ward's thesis on education and achievement is echoed 66 years later by A. R. Jensen: "We have to face it: the assortment of persons into occupational roles simply is not 'fair' in any absolute sense. The best we can hope for is that true merit, given equality of opportunity, acts as a basis for the natural assorting process"(5; p. 15).

The ideology of the modern competitive market society is then not one of equality of station but one of a natural sorting process aided by universal education in which "intrinsic merit" will be the criterion and source of success. The social program of the state, then, should not be toward an "unnatural" equalization of condition, which in any case would be impossible because of its "artificiality," but rather the state should provide the lubricant to ease and promote the movement of individuals into the positions to which their intrinsic natures have predisposed them.

The concept that social arrangements are a manifestation of the inner or intrinsic natures of human beings and are therefore unchangeable has come to be called *biological determinism*. As we shall see, the degree of rigidity of the determinism varies in different versions of the system, from the notion that biological factors virtually determine completely the "nature" of each individual to the more subtle idea that human biological nature establishes only "tendencies," natural states toward which human beings will gravitate in the normal course of events. Biological determinism has two complementary facets, both of which are necessary to complete this scheme. First, it is asserted that the *differences* in manifest abilities and power between individuals, classes, sexes, races, and nations result in large part from differences in intrinsic biological properties of individuals. Some of us can paint pictures and others can only paint houses (Jensen[5]), while some of us can be doctors but others can only be barbers (Herrnstein[6]). But these facts alone, if they were true, would not in themselves necessarily result in a society of unequal power. After all, there is no reason that differences in ability, whether intrinsic or not, need imply differences in status, wealth, and power. We might build a society in which picture painters and house painters, barbers and surgeons would be given equal material and psychic rewards. This is the argument of Dobzhansky in *Genetic Diversity and Human Equality*(7). If taken seriously, this argument would deprive our unequal society of legitimacy offered to it by the argument of biological diversity. To complete its function as a legitimation argument for the present state of the world, biological determinism requires a second facet, the belief in *human nature*. In addition to the biological differences between individuals and groups, it is supposed that there are biological "tendencies" shared by all human beings and their societies and that these tendencies result in hierarchically organized societies in which individuals "compete for the limited resources allocated to their role sector. The best and most entrepreneurial of the role-actors usually gain a disproportionate share of the rewards, while the least successful are displaced to other, less desirable positions"(8; p. 554).

The assertion that "human nature" guarantees that the biological differences among individuals and groups will be translated into differences in status, wealth, and power is the other face of biological determinism as a total ideology and represents the consolidation phase of the bourgeois revolution. To justify their original ascent to power, the new middle class had to demand a society in which "intrinsic merit" *could be* rewarded. To maintain their position of power, they claim that intrinsic merit, once free to assert itself, *must be* rewarded. It is all natural and inevitable, so why fight it?

One element is left to complete the ideology and bring it to perfection as a weapon in social warfare. It is easily observed that even in a democratic society rewards are not reassorted each generation. The children of oil magnates tend to become bankers, while the children of oil workers tend to be in debt to banks. Can it be that parents are passing their social power to their children and thus circumventing the perfect assortative process based on intrinsic merit? Hardly. It must be that the biological abilities that are rewarded are passed on biologically from parent to child. Thus, we have the equation of biological differences with hereditary differences that assures a legitimate passage of social position from generation to generation. The equation of *biological* with *hereditary* is clearly not essential logically, since inborn differences might easily arise from accidents of development. Folklore reflects an appreciation of this possibility in the notion that the physical and psychic dispositions of children may be influenced by experiences of their mothers during pregnancy. It is not clear when the equation of *biological* with *hereditary* became common, but it certainly predates modern genetics. Nineteenth-century literature is permeated with the notion that human behavior is inherited. The classic expression is in Zola's Rougon-Maquart novels, which chronicle the two halves of the same family, descendants of one woman by two men. The descendants of the husband Rougon, a solid, hard-working peasant, are intelligent, hard working, and ambitious, while those who sprang from the dissolute, drunken, criminal lover Maquart are equally degenerate and alcoholic. Among the Maquarts are Gervaise, the hard-working, successful laundress who nevertheless finally succumbs to her inherited laziness and drunkenness, and her daughter Nana, sexually degenerate from early childhood. The Rougon-Maquarts are the type for the American myth of the Kallikaks, which has graced textbooks of psychology for years (for example, Garrett's *General Psychology* [9]). Martin Kallikak, a colonial soldier, had two wives, one half-witted and dissolute and the other respectable and middle class, and the respective branches of the family followed the type to remote generations. Thus, Kallikak's descendants through his middle-class wife are all good, solid citizens, while those through his other wife are shameful degenerates.

English literature, too, has demonstrated the rule of nature over nurture. Oliver Twist, raised from birth in that most degrading social institution—the parish workhouse—and educated in crime by Fagin, nevertheless develops gentleness, honesty, and the Christian virtues and all the while speaks perfect, grammatical English. All is explained when it turns out that he is the child of a respectable upper-middle-class woman. The most remarkable case is George Eliot's Daniel Deronda, who is raised from birth by an English nobleman and develops into the typical leisure-class, gambling dandy of the nineteenth century but who in early adulthood feels mysterious longings and attractions for things Hebrew, including a passion for a Jewish woman. The reader will not be surprised to learn that he is really the son of a Jewish actress.

In the twentieth century, modern genetic ideas have replaced the vague notions of "blood," but nothing else has changed. Oliver Twist and Daniel Deronda are the prototypes of the modern adoption study, but Dickens and Eliot were better experimenters than their modern counterparts who have failed to transgress class lines in their baby exchanges (see the Schwartz article in this volume). Only in the imagination of a Victorian novelist or a Gilbert and Sullivan plot can children be distributed at random across social boundaries from an early age. The rediscovery of Mendel in 1900 very quickly provided a scientific apparatus that could be marshaled to produce "scientific" explanations and an apparatus of objectivity to support the claims of hereditarians for the supremacy of innate factors. Thus, E. L. Thorndike, characterized by A. R. Jensen as "probably America's greatest psychologist and a pioneer in twin studies of the heritability of intelligence"(10; p. 17), wrote in a scientific paper on twins that "in the actual race of life, which is not to get ahead, but to get ahead of somebody, the chief determining factor is heredity"(11; p. 12). This assertion that the "chief determining factor is heredity" was made in 1905, only 5 years after the rediscovery of Mendel's paper, but 13 years *before* Fisher's paper establishing the statistical theory on which genetic studies of quantitative characters are based, 10 years *before* Fisher's derivation of the sampling distribution of the correlation coefficient, and 5 years *before* Morgan's chromosome theory of inheritance. E. L. Thorndike appears to have been not only America's greatest psychologist but its greatest geneticist, statistician, and crystal-ball gazer as well. And he was not an exception. America's most prestigious academics and scholars in psychology, sociology, and biology have over and over again asserted as *facts* what they cannot have known to be true. They have used their immense authority to misinterpret, misinform, and sometimes deliberately misrepresent biological concepts and observations in the service of an ideology to which they adhere.

THE BASIC FALLACY

The assertion that individuals, sexes, races, classes, and nations owe their conditions to inherited qualities rests on a conceptual error of the most fundamental kind. This error is to believe that what is inherited is a "tendency" or a "capacity" that is then somehow modulated to a greater or lesser degree by environment. This essentialist notion postulates a particular morphology, physiology, or behavior as the ideal or "true" innate characteristic of a given genetic constitution, while the realized organism is an imperfect manifestation of this Platonic ideal. Thus, Herrnstein, quoting L. M. Terman, the founder of the American psychological testing movement and the inventor of the Stanford-Binet IQ test, speaks of the "extent to which IQs can be artificially raised"(6; p. 54). But to raise IQs *artificially* implies that there is a *natural* level of IQ appropriate to each person. A related notion is that genes specify an individual's "capacity." This is the empty-bucket metaphor: each of us at birth is an empty bucket that will be filled by experience and education. Some of us have large buckets and some small, so that, no matter how much is poured into them, some will eventually have much more than others. Education is wasted on the small buckets, and greater education and a richer environment can only make the difference in final outcome greater. Genetic differences will become more manifest in better environments.

But there is nothing in genetics to support the concept of a Platonic ideal that is the preferred state of a genotype. Genetics has nothing to say about "capacities," and the empty-bucket metaphor is a complete falsification of our knowledge of gene action. What the genotype does specify is a pattern of reaction of a developing organism to the sequence of environments encountered by the organism. For each environmental sequence, there will be a unique developmental result, and this result varies from genotype to genotype. There is no particular outcome of the development that is characteristic of a given genotype that is more "natural" or "typical" of the genotype than any other. We cannot speak of "artificially" raising IQs. If genes have any influence in normal cognitive development, that influence can only be expressed by the *norm of reaction,* the table of correspondence between the set of environments and the set of developmental outcomes. A particular genotype may grow tall in one environment and short in another, while another genotype behaves in the opposite way in the same two environments. There is absolutely no evidence on what the pattern of reaction of human genotypes is over environments, because the necessary experiments cannot be carried out. To trace out a human norm of reaction for any trait, it would be necessary to produce large numbers of genetically identical individuals, perhaps by cloning, and then to place the developing individuals in a variety of different social and individual environments. In the absence of such a science-fiction experiment, we do not know what human reaction norms look like, but we do know about the norms of a variety of experimental animals and plants for a variety of morphological, physiological, and behavioral characteristics over a variety of environments. The overwhelming evidence is that there are no unconditionally "superior" genotypes that outperform other genotypes over the whole range of tested environments. On the contrary, the typical result is that, when genotype A, for example, has the highest survival rate of all the tested genotypes in one environment, it will be mediocre in other environments. When inbred lines of rats are tested for passive avoidance learning, one strain may learn faster at a low stimulation level, but a second strain may be superior at a high stimulation level. It is important to note that it is *not* being asserted that there are no genetic differences in such experiments but that these genetic differences cannot be described as a difference in "capacity" or "tendency," a description that implies that there is one value that can be isolated as the true or natural characteristic of each genotype.

From the fallacy of a true or natural tendency of each genotype, there also flows the fallacy of human nature. According to the human nature doctrine, despite manifest variations in human cultures, there is a tendency for all human cultures to have a similar structure and content. "Man," the argument runs, "tends to be entrepreneurial, aggressive, territorial, curious, et cetera, et cetera." If individuals or cultures do not clearly manifest such characteristics, they are thought to be in a state of temporary aberration or else to express the tendencies in a disguised or sublimated form.

A second incorrect consequence that is drawn from the erroneous typological view of genotypes is that characters influenced by genes are difficult to change. A. R. Jensen's famous article on IQ was entitled "How Much Can We Boost IQ and Scholastic Achievement?" and the answer given was "not much" because IQs are mostly inherited. But this is a *non sequitur.* There is no relationship, either logical or empirical, between the social and environmental plasticity of a characteristic and the presence or absence of genetic differences among individuals. In one environment, there may be immense differences between different genotypes in their development; yet, a change in environment may cause an arbitrarily large alteration in the character, and the genotypes may become indistinguishable from each other. A knowledge of the role of genes in influencing IQ or any other trait gives us no information on the plasticity of the trait. To counterpose nature versus nurture, heritability versus plasticity, is a total distortion of physical reality in the interest of "proving" that the *status quo* is natural and therefore either unchangeable or changeable only at the expense of a constant totalitarian control forcing humans beyond their "natural" behavior.

THE FORMS OF DETERMINISM

There have been four major social forms taken by biological determinism, each supported by a different but sometimes overlapping group of academic ideologies. These four are *racism, class superiority, sexism,* and *human nature.*

From the nineteenth century to the present, there have been repeated claims by eminent academics that scientific facts point clearly and objectively to the superiority of one race over another and one set of ethnic groups over others. Not surprisingly, the academic science produced by white northern European culture has consistently shown the racial superiority of white northern Europeans. This alleged superiority is always based on objective scientific evidence, of course. So, America's most distinguished zoologist of the nineteenth century, Louis Agassiz, a professor of zoology at Harvard, wrote that "we have a right to consider the questions growing out of men's physical relations as merely scientific questions, and to investigate them without reference to politics or religion"(12; p. 110). The sentiment was echoed in 1975 by another Harvard professor, Bernard Davis, who assures us that "neither religious nor political fervor can command the laws of nature"(13). But political fervor apparently can command what professors *say* about the laws of nature since, as Stanton points out, Professor Agassiz claimed that "the brain of the Negro is that of the imperfect brain of a seven months infant in the womb of the white"(14; p. 106) and that the skull sutures of black babies closed earlier than those of whites, so that it was dangerous to teach black children too much lest their brains swell beyond the capacity of their skulls and burst!

Agassiz was not a lone, backward, nineteenth-century crackpot. No finer example of the marshaling of scientific facts in a logically impeccable argument can be found than in the remarkable claim of Henry Fairfield Osborn, president of the American Museum of Natural History and one of America's most eminent and prestigious paleontologists, who worked out the evolution of the horse, that(15):

The northern races . . . invaded the countries to the south, not only as conquerors, but as contributors of strong moral and intellectual elements to more or less decadent civilizations. Through the Nordic tide which flowed into Italy came the ancestors of Raphael, Leonardo, Galileo, Titian; also, according to Günther, of Giotto, Botticelli, Petrarch and Tasso. Columbus, from his portraits and from busts, whether authentic or not, *was clearly of Nordic ancestry. Kosuth was a Calvinist and of a noble family and there is presumption in favor of his being Nordic. Kosciusko and Pulaski were members of a Polish nobility which, at that time, was largely Nordic. . . . [My emphasis.]*

Osborn was one of the major supporters of the American eugenics movement, the purpose of which was to prevent racial degeneration from an influx of the foreign-born and blacks. It is now quite clear that biologists and psychologists who supported this movement played a major role in providing intellectual legitimacy for the Immigration Act of 1924, which created ethnic quotas for immigrants strongly weighted toward northern Europeans(16). The founder of the eugenics movement, Sir Francis Galton, had wondered that "there exists a sentiment, for the most part quite unreasonable, against the gradual extinction of an inferior race"(17; p. 200[11]). There is nothing to indicate that his American followers in any way moderated this view. In 1916, L. M. Terman found that an IQ between 70 and 80 is "very common among Spanish, Indian and Mexican families . . . and also among Negroes. Their dullness seems to be racial, or at least inherent in the family stocks from which they come . . . from the eugenic point of view they constitute a grave problem because of their prolific breeding"(18; pp. 91-92). Henry Garrett, chairman of the psychology department at Columbia for 15 years and president of the American Psychological Association, generalized that "whenever there has been mixed breeding with the Negro there has been deterioration in civilization"(19).

In 1923, when he taught at Princeton, Carl Brigham, secretary of the College Entrance Examination Board, produced *A Study of American Intelligence* under the direction of R. M. Yerkes, professor of psychology at Harvard and another president of the American Psychological Association. According to that study: "We must assume that we are measuring inborn intelligence. . . . We must face the possibility of racial admixture here that is infinitely worse than that faced by any European country, for we are incorporating the negro into our racial stock. The decline of American intelligence will be more rapid . . . owing to the presence here of the negro"(20; pp. 209-10).

And so the dreary litany goes on. Presidents of the American Psychological

Association and professors at America's leading universities—objective scientists all—continue to assert that whites are genetically superior to blacks, northern Europeans to degenerate Slavs and Mediterraneans, in defiance of all facts and logic.

The supposed biological inferiority of dark races and southern nationalities has been coupled with and has merged into a claim that lower social classes are also biologically inferior. Since blacks, southern and eastern European immigrants, and Latin Americans have made up a disproportionate share of the exploited class in the United States, it is not surprising that this connection of race and class has been made. It reflects the manifest reality. Herrnstein's article on IQ and social class was introduced by the editors of *The Atlantic Monthly* in a short essay specifically linking it to "three landmark documents—by Daniel Patrick Moynihan, James Coleman and Arthur R. Jensen"(6), all dealing with the situation of blacks. Moreover, in the article itself, Herrnstein claims to have explained the troubles caused by "the increasingly chronic lower class in America's central cities." Of course, everyone knows who lives in America's "central cities" and what sort of trouble they have caused. Nevertheless, Herrnstein's claims are chiefly devoted to lower socioeconomic classes irrespective of color and ethnicity, and he makes the same error for social class that Jensen makes for race—the error that the existence of genetic differences among individuals is somehow evidence for a genetic difference among groups. The assertion that the lower socioeconomic classes are genetically inferior goes back to the founders of the mental testing movement. Terman wrote, "If we are to preserve our state for a class of people worthy to possess it, we must prevent, as far as possible, the propagation of mental degenerates, . . . curtailing the increasing spawn of degeneracy"(21). The leading American idealogue of the innate mental inferiority of the working class was, however, H. H. Goddard, a pioneer of the mental testing movement, the discoverer of the Kallikak family, and the administrant of IQ tests to immigrants that found 83 percent of the Jews, 80 percent of the Hungarians, 79 percent of the Italians, and 87 percent of the Russians to be feebleminded(16). Goddard was worried in 1917: "the disturbing fear is that the masses—the 70 or even 86 million—will take matters into their own hands. The 4 million or so, of superior intelligence, must guide and direct the masses"(22; p. 97). Moreover, Goddard was concerned lest the "4 million or so, of superior intelligence," assume that the masses were the victims of some form of social injustice rather than that they were innately defective. Such an erroneous assumption might lead to socialist ideas even among Princeton undergraduates(22; pp. 100-103):

> *These men in their ultra altruistic and humane attitude, their desire to be fair to the workman, maintain that the great inequalities in social life are wrong and unjust. . . . As we have said, the argument is fallacious. It assumes that [the] laborer is on the same mental level as the man defending him.*
>
> *Now the fact is [the] workmen may have a ten year intelligence while you have a twenty. To demand for him such a home as you enjoy is as absurd as it would be to insist that every laborer should receive a graduate fellowship. How can there be such a thing as social equality with this wide range of mental capacity? . . . As for an equal distribution of wealth of the world, that is equally absurd. . . . These facts are appreciated. But it is not fully appreciated that the cause is to be found in the fixed character of mental levels. In our ignorance we have said, let us give these people one more chance—always one more chance.*

Seldom has the weapon of biological determinism been so blatantly brandished in an effort to maintain the existing state of society.

It is probably a manifestation of the largely unquestioned role of women in our society that the heavy-calibre weapons in the hands of the most prestigious biologists and psychologists were not directed, for a long time, against the equality of the sexes.

If Terman, Yerkes, Osborn, Agassiz, and the others felt as threatened by women as they did by blacks, immigrants, and the working class, they did not manifest their fears in their major pronouncements. Even now, despite the growing women's movement, the number of academics who are willing to publish and legitimize the attitudes they express in private is small, but a few have, and there is some evidence that even the most prestigious are about to enter the fray. The claims of Tiger and Fox(23) for the biological superiority of men were a well-known feature of "pop" ethology a few years ago, and a similar vein of vulgarized science is contained in Goldberg's *The Inevitability of Patriarchy,* which makes the claim that "human biology precludes the possibility of a human social system whose authority structure is not dominated by males, and in which male aggression is not manifested in dominance and attainment of position, of status and power"(24; p. 78).

We are not told how the discoveries of biology "preclude the possibility" of female equality or domination, but it is clear from the work as a whole that the author believes that "tendencies" inherent in males and females lead ineluctably to a "naturally" asymmetric social system. In addition to their innately greater aggression, "The stereotype that sees the male as more logical than the female is unquestionably correct in observation, and probably correct in its assumption that the qualities observed conform to *innate sexual limitations* analogous to those relevant to physical strength"(24; p. 204). [My emphasis.] The two strains of aggressivity and logic are explicitly drawn together by Eleanor Maccoby, who suggests that "there is good reason to believe that boys are innately more aggressive than girls . . . and if this quality is one which underlies the later growth of analytic thinking, then boys have an advantage which girls will find difficult to overcome"(25; p. 37). Like Goldberg, Maccoby brings in fallacious notions of innate tendencies and then converts these tendencies into limitations on groups. The entire typology of "*the* male as more logical than *the* female" is an outmoded nineteenth-century concept of typical individuals standing for entire groups. What proportion of males manifest a greater logical ability than what proportion of females? What are the "innate" differences in population means? Is the "tendency" manifest simply as a small difference in the average of all males as opposed to the average of all females? If so, why does a difference in average "preclude the possibility" of eliminating the dominance of women by men. The intellectual bankruptcy of the vague speculations of male intrinsic superiority immediately appears when any attempt at analysis is made.

The reader should not imagine that the inevitability of male domination is a feature only of the writings of popularizers. The most recent declaration of the biologically inevitable domination of women by men has been made by E. O. Wilson, professor of zoology at Harvard, a man generally regarded as a leading authority on the evolution of animal social behavior: "In hunter-gatherer societies, men hunt and women stay at home. This strong bias persists in most agricultural and industrial societies and, on that ground alone, appears to have a genetic origin. . . . My own guess is that the genetic bias is intense enough to cause a substantial division of labor even in the most free and most egalitarian of future societies. . . . Even with identical education and equal access to all professions, men are likely to continue to play a disproportionate role in political life, business and science"(26; pp. 48 and 50).

The theory that the relation of domination of men over women that characterizes our society has a biological cause and is thus inevitable provides a bridge between theories that differences between groups are genetic and theories that human societies are the result of an innate "human nature." Ideas of human nature appear in a great diversity of social theories and in each one explicitly serve to legitimize political ends. Not even the historicist argument of Marx and Engels was free of an occasional appeal to human nature, in their case to unalienated labor as the essence of human self-realization(27). Like the claims of the natural inferiority of women, recent arguments about the true nature of man have been mostly in the popularizations of science, like those of Ardrey(28) and Tiger and Fox, in which it is argued that the human species is naturally territorial, aggressive, male dominant, and so forth, and carefully selected observations are used from ethnography, paleontology, and animal behavior. Konrad Lorenz, Nobel Prize winner in ethology, has attempted to give human relevance to his observations on lower animals in *On Aggression*(29). He argues that humans lack the built-in controls against intraspecific aggression that characterize other dangerous animals, because, during most of our evolution, we were not predatory carnivores, and, therefore, some social control of natural human aggression and nastiness must be exercised. More important, the domestication of man has resulted in the loss of natural tendencies to reject from the species "degenerate" types. This rejection then also must be exercised by a social agency. In particular, Lorenz wrote in *1940 in Germany* during the Nazi extermination campaign: "The selection of toughness, heroism, social utility . . . must be accomplished by some human institution if mankind, in default of selective factors, is not to be ruined by domestication induced degeneracy. The racial ideal as the basis of the state has already accomplished much in this respect"(30; p. 71).

Human-nature ideologies are based firmly in the belief that human behavioral attributes of all kinds are genetically determined. That belief, with no shred of experimental evidence to back it up, has been widespread among geneticists who have simply assumed that everything they see *must* be genetic. Perhaps the greatest unintentional caricature of this arch-hereditarian position is in a list of traits to be used in choosing sperm donors for an improvement of the human species. The author of the list is H. J. Muller, a Nobel laureate who is generally agreed to be the greatest geneticist after T. H. Morgan. Some of the traits on Muller's list include "joy of life, strong feelings combined with emotional self-control and balance, the humility to be corrected and self-corrected without rancor, empathy, thrill at beholding and at serving in a greater cause than one's own self-interest, fortitude, patience, resilience, perceptivity, sensitivities and gifts of musical and other artistic types, expressivity, curiosity, love of problem solving. The list is very incomplete"(31; p. 535).

Muller does not give us the literature citation for the study that established the heritability of "thrill at beholding and at serving in a greater cause than one's own interest." Nor was Muller qualitatively atypical of geneticists, although he did represent a quantitative extreme.

The newest wave of human nature determinism has culminated in the publication by E. O. Wilson of *Sociobiology: The New Synthesis*(8), which announces the creation of a new field, sociobiology, and which asserts that such human cultural manifestations as religion, ethics, tribalism, warfare, genocide, cooperation, competition, entrepreneurship, conformity, indoctrinability, and spite (the list is incomplete) are tendencies that have been established by natural selection and by implication are encoded in the human genome. No evidence at all is presented for a genetic basis of these characteristics, and the arguments for their establishment by natural selection cannot be tested, since such agruments postulate hypothetical situations in human prehistory that are uncheckable. For example, homosexuality is hypothesized to be genetically conditioned (no evidence), and it is then assumed that homosexuals leave fewer offspring than heterosexuals (no evidence and a confusion between homosexual *acts* and total homosexuality), but then it is postulated that the "genes" for homosexuality may have been preserved in human prehistory because homosexuals served as helpers to their close relatives (uncheckable story with no ethnographic evidence from present hunters and gatherers to suggest such a phenomenon). The intended use of sociobiology in human social affairs is made crystal clear by its inventor, however. The book begins on page 4 with the statement that "it may not be too much to say that sociology and the other social sciences, as well as the humanities, are the last branches of biology waiting to be included in the Modern Synthesis. One of the functions of sociobiology, then, is to reformulate the foundations of the social sciences in a way that draws these subjects into the Modern Synthesis." And the book ends on page 574 with a vision of neurobiologists and sociobiologists as the technocrats of the near future who will provide the necessary knowledge for ethical and political decisions in the planned society: "If the decision is taken to mold cultures to fit the requirements of the ecological steady state, some behaviors can be altered experientially without emotional damage or loss in creativity. Others cannot. Uncertainty in the matter means that Skinner's dream of a culture predesigned for happiness will surely have to wait for the new neurobiology. A genetically accurate and hence [sic] completely fair code of ethics must also wait."

Sociobiology is basically a political science whose results may be used, eventually, as the scientific tools of "correct" social organization. Yet, the world to be made will be pretty much the aggressive, domination-ridden society we live in now. Why is that? Because, as Wilson says on page 575:

We do not know how many of the most valued qualities are linked genetically to more obsolete destructive ones. Cooperativeness toward groupmates might be coupled with aggressivity toward strangers, creativeness with a desire to own and dominate, athletic zeal with a tendency to violent response, and so on. . . . If the planned society—the creation of which seems inevitable in the coming century—were to deliberately steer its members past those stresses and conflicts that once gave the destructive phenotypes their Darwinian edge, the other phenotypes might dwindle with them. In this, the ultimate genetic sense, social control would rob man of his humanity.

Of course, it is all put in a hypothetical mode, but the message is clear: the only safe thing to do is to leave things as they are, at least at present. Don't rock the boat until the sociobiologists tell you how.

THE SCHOOLS

I have tried to establish that the most prestigious academics in the most prestigious academic institutions have over and over again tried to legitimize a given social order by the production of ideological tools. But the real struggle for the minds and consciousness of human beings is not in the academy. It is, in large part, in the schools. The institutions of higher learning and research are only the weapons factories in the social class war. The schools are the battlefields on which those weapons are used with devastating force to paralyze the wills and destroy the reasons of those who are most victimized by the system of domination in which they will live. Does that seem an absurdly radical and extremist view of the educational system? Then, let the school textbooks and the theorists and the builders of the school system speak for themselves.

Frank Freeman, professor at the University of Chicago and leading sociologist of education: "It is the business of the school to help the child acquire such an attitude toward the inequalities of life, whether in accomplishment or reward, that he may adjust himself to its conditions with the least possible friction"(32; p. 170). A second-grade reader (Senesh's *Our Working World*[33; pp. 177-78]):

What did we learn?

1. *We go to school to learn about people.*
2. *We learn that people look different from each other.*
3. *People speak different languages.*
4. *Some people have more than others, some have less.*
5. *Some people know more than others, some know less.*
6. *Some people can learn a lot. Some people can learn only a little.*

And last, but not least, Daniel Webster: "Education is a wise and liberal form of police by which property and life and peace of society are secured."

References

1. Marx, K., and Engels, F. 1847. *Manifesto of the communist party.* New York: International Publishers (1948).
2. Herrnstein, R. J. 1973. *I.Q. in the meritocracy.* Boston: Atlantic-Little, Brown and Company.
3. Ward, L. F. Education. 1873. An unpublished manuscript available in the Special Collection Division, Brown University, Providence, R.I.
4. Ward, L. F. 1903. *Pure sociology.* New York: Macmillan.
5. Jensen, A. R. 1969. How much can we boost IQ and scholastic achievement? *Harvard Educational Review* 39:1-123.
6. Herrnstein, R. 1971. I.Q. *The Atlantic Monthly* 228(3):43-64.
7. Dobzhansky, T. 1973. *Genetic diversity and human equality.* New York: Basic Books.
8. Wilson, E. O. 1975. *Sociobiology: The new synthesis.* Cambridge, Mass.: Harvard University Press.
9. Garrett, H. E. 1955. *General psychology.* New York: American Book Company.
10. Jensen, A. R. 1970. Race and the genetics of intelligence: Reply to Lewontin. *Bulletin of the Atomic Scientists* 26(5):17-23.
11. Thorndike, E. L. 1905. Measurement of twins. *Archives of Philosophy, Psychology, and Scientific Methods,* vol. 1, pp. 1-65. Eds. J. McKeen Cattell and J. E. Woodbridge. In *Columbia University Contributions to Philosophy and Psychology,* vol. 8, no. 3. New York: Science Press.
13. Davis, B. 1975. Social determinism and behavioral genetics. *Science* 189:1049.
14. Stanton, W. 1960. *The leopard's spots: Scientific attitudes toward race in America 1815-1859.* Chicago: University of Chicago Press.
15. Osborn, H. F. 1924. Letter to *The New York Times. The New York Times,* April 8, 1924, p. 18.
16. Kamin, L. 1974. *The science and politics of I.Q.* New York: Halsted Press. [See also, Allen, G. E. 1975. Genetics, eugenics and class struggle. *Genetics* 79 (supplement):29-45.]
17. Galton, F. (no date) *Inquiries into human faculty and its development.* 2nd ed. New York: E. P. Dutton and Company. [Quoted in Haller, M. H. 1963. *Eugenics: Hereditarian attitudes in American thought.* New Brunswick, N.J.: Rutgers University Press.]
18. Terman, L. M. 1916. *The measurement of intelligence.* Boston: Houghton Mifflin Company.
19. Garrett, H. L. (no date) *Breeding down.* Richmond, Va.: Patrick Henry Press.
20. Brigham, C. C. 1923. *A study of American intelligence.* Princeton, N.J.: Princeton University Press.
21. Terman, L. M. 1917. Feeble-minded children in the public schools of California. *School and Society* 5:165.
22. Goddard, H. H. 1920. *Human efficiency and levels of intelligence.* Princeton, N.J.: Princeton University Press.
23. Tiger, L., and Fox, R. 1970. *The imperial animal.* New York: Holt, Rinehart and Winston.
24. Goldberg, S. 1973. *The inevitability of patriarchy.* New York: William Morrow Company.
25. Maccoby, E. 1963. Woman's intellect. In *Man and civilization: The potential of women,* eds. S. Farber and R. H. L. Wilson. New York: McGraw-Hill.
26. Wilson, E. O. 1975. Human decency is animal. *New York Times Magazine,* October 12, 1975, pp. 38-50.
27. Engels, F. 1934. The part played by labor in the transition from ape to man. In *Dialectics of nature.* Moscow: Progress Publishers.
28. Ardrey, R. 1966. *The territorial imperative.* New York: Atheneum.
29. Lorenz, K. 1966. *On aggression.* New York: Harcourt Brace Jovanovich.
30. Lorenz, K. 1940. Durch Domestikation verursachte Störungen arteigenen Verhaltens. *Zeitschrift für angewandte Psychologie und Characterkunde* 59:2-81.

31. Muller, H. J. 1961. What genetic course man steers. *Proceedings of the Third International Congress of Human Genetics,* eds. J. F. Crow and J. V. Neel. Baltimore: Johns Hopkins University Press.
32. Freeman, F. 1924. Sorting the students. *Educational Review,* November 1924, p. 170.
33. Senesh, L. 1965. *Our working world.* Chicago: Science Research Associates.

Race and IQ

The race and intelligence controversy is a very old one, but its most recent reemergence into public life began in 1969 with the publication of Arthur Jensen's "How Much Can We Boost IQ and Scholastic Achievement?" in the *Harvard Educational Review*(1). Jensen's long paper, which began "Compensatory education has been tried and it apparently has failed," argued that differences in IQ between races are due mostly to genetic factors, so that it is useless to try to eliminate them through special educational programs such as Head Start. Jensen based his argument on the concept of the heritability of IQ and claimed that it had been shown to be high (60-80 percent) in many studies and that this figure, while generally measured within a single race (whites), tells us how much of the difference between races is due to genes.

Jensen's article was widely publicized: it was sent to every member of the National Academy of Sciences, was read into the *Congressional Record,* and later was expanded into a book(2). The IQ argument was also put forward by other authors, with some variations. Richard Herrnstein applied it to social classes, arguing that success in our society derives from high IQ and that poor people are limited in IQ by their genetic potential(3). Thus, class differences were inevitable. Indeed, as educational opportunities were equalized, social classes would become more and more clearly differentiated genetically. In Britain, Henry Eysenck repeated the IQ argument about racial inferiority(4). And William Shockley used it as the basis for proposing a plan for sterilization of individuals with low IQs(5).

Among the many critiques of the IQ argument in recent years, three seem to stand out. Richard Lewontin pointed out in the *Bulletin of the Atomic Scientists* that heritability measured *within* a particular group (such as a race) tells nothing about the differences *between* that group and others(6). Leon Kamin, in *The Science and Politics of I.Q.*, reviewed the original evidence on which the high estimates of heritability are based and found much of it shaky and possibly even falsified(7). Finally, Samuel Bowles, Valerie Nelson, and Herbert Gintis have cast doubt on the basic assumption of nearly all participants in the debate: that IQ is an important determinant of success in our society(8; pp. 30-35 and 102-24).

The practical importance of the IQ debate lies in the use of IQ scores to determine the type of education given to different people, a fact which was recognized as long ago as 1924, when the Chicago Federation of Labor attacked the use of IQ tests to assign students to various "tracks" in the public schools(8; p. 195). Beyond this, the IQ debate has important effects on one's view of race and class differences in our society. Thus, it promises to remain an underlying issue in political debate long after the current controversy fades away.

References

1. Jensen, A. R. 1969. How much can we boost IQ and scholastic achievement? *Harvard Educational Review* 39:1-123.
2. Jensen, A. R. 1973. *Genetics and education.* New York: Harper and Row.
3. Herrnstein, R. J. 1973. *I.Q. in the meritocracy.* Boston: Atlantic-Little, Brown and Company.
4. Eysenck, H. J. 1971. *The IQ argument: Race, intelligence, and education.* New York: Library Press.
5. Shockley, W. 1972. Dysgenics, geneticity, and raceology. *Phi Delta Kappan,* January 1972, p. 305.
6. Lewontin, R. C. 1970. Race and intelligence. *Bulletin of Atomic Scientists* 26:2-8.
7. Kamin, L. 1974. *The science and politics of I.Q.* New York: Halsted Press.
8. Bowles, S., and Gintis, H. 1975. *Schooling in capitalist America.* New York: Basic Books.

The Politics of Statistics: Heredity and IQ

Arthur J. Schwartz

Throughout the reign of the Stuarts, the divine right of kings was proven by their ability to cure scrofula (lymphatic tuberculosis, also known as the king's evil). We may be skeptical, yet this ability was attested to by many of high rank, integrity, and probity. In recent years, this regal curative power has faded. Instead, the modern method for detecting hereditary superiority is the IQ test, and many of high rank, integrity, and probity attest to its value. The loudest voices supporting the social significance of IQ scores are those of Jensen, Shockley, Herrnstein, and Eysenck. They argue that a major component of intelligence is genetically determined. The effect of their efforts is to undermine attempts to eliminate inequities in education, employment, and the quality of life.

It is the purpose of this article to show that there is no foundation, conceptual or empirical, to the hereditarian position. First, we shall consider the concept of intelligence and its measurement, the philosophy and politics of the testing movement, and how these influence the tests themselves. Second, we shall deal with the historical background of the theory of human statistics, its application to the intelligence/heredity nexus, and the actual practice of collecting and analyzing data.

INTELLIGENCE AND ITS MEASURERS

A Definition of Intelligence

To begin with, what do we mean by intelligence? Here is a definition given by Loehlin, Lindzey, and Spuhler in *Race Differences in Intelligence* (1; p. 49):

A person walks into a situation in which others are floundering, appraises it, and selects an effective course of action. If he does such a thing only once, we may say he is lucky. If he can do it only in particular kinds of situations, we may say he has a special knack or talent. But if he does it over and over again, in a wide variety of situations with which his prior familiarity is no greater than yours or mine, we say he is intelligent.

This seems to be the core, common sense meaning of the term intelligence. *We may paraphrase it as "general problem-solving ability," if we recognize that the "problems" comprise a very wide range of situations, and that abilities can be recognized by their fruits in word or deed.*

In brief, then, we consider an act to be intelligent if it reflects a grasp of the essential features of an obscure or complex situation; we consider a person to be intelligent if he characteristically performs intelligent acts; and we consider intelligence (if we elect this degree of abstraction) to be the attribute of the person that enables him to behave in this fashion.

One would indeed be fortunate to have such a universally applicable ability. It would also be nice to know a woman or a man with such excellent characteristics. Unfortunately, I have never met such a person, and I doubt that many people have. Moreover, a reliable test for this all-around problem-solving ability would surely provide first-class material for the Brothers Grimm.

Certainly, there is great diversity in intellectual makeup. Each of us, no doubt, tends to appraise acquaintances differently. A person may seem to us stimulating or boring, understanding or callous, knowledgeable or ignorant (of various subjects), perceptive or insensitive, handy or helpless, quick or slow, and so on. Within each of these categories, there are subcategories and subgradations. But if I make an intellectual appraisal of someone, there is no guarantee that it will agree with anyone else's appraisal of that person nor is there any reason that it should. Intelligence is a multidimensional, interpersonal, interactive, subjective phenomenon.

In other parts of their book, Loehlin, Lindzey, and Spuhler do allow that intelligence may be complex and multidimensional, but, as is the case with many authors who deal with this subject, they soon revert to discussing that which is measured by IQ tests. Thus, they begin with a heroic, one-dimensional ideal, they then mention the possibility that things are not quite so simple, and, ultimately, they turn to a consideration of tests that measure something far less significant. Unfortunately, long before one reaches the end of such expositions, all caveats and expressions of doubts about the relationship between IQ measurements and what we may intuitively understand as intelligence have disappeared from view.

Mental Tests

Perhaps, this natural skepticism about the possibility of assigning a numerical value to intelligence in some precise and meaningful way can be overcome by the presentation of some remarkable testing procedure. Supposedly, the testers and quantifiers have in fact developed such instruments with which human intelligence can be measured. These are the so-called mental or IQ tests. Reading through a number of different examples of these tests, I was struck by features common to all of them.

First, none of the questions or problems seem incisive. They are, in fact, very bland. There is no reason to believe such problems can probe a subject's mind; on the contrary, their major effect may be to bore the test taker. Presenting a number of such problems does not make the test more incisive, but it certainly increases the risk of boredom. To some extent, we may be measuring an individual's ability to tolerate being bored.

Second, in many cases, there is more than one reasonable answer or solution to a problem, although only one is accepted as correct by the examiner. The following are examples taken from R. Pintner's *Intelligence Testing, Methods and Results*(2; pp. 183-89) (Pintner, who has designed well-known tests, indicates no concern for the ambiguity in the examples below):

Underline the word in parentheses which is the opposite of the first word:

accept . . . (receive, percept, deny, reject, spend)
constant . . . (always, fickle, stationary, seldom, movable)

Underline the extra word:

. . .

clarify, explain, argue, illuminate, elucidate

?????

It is likely that most readers of the present article are fairly good test takers and will agree with my selection of *reject* and *fickle* as the *desired* answers for the first part. Nevertheless, the opposite of *accepting* an idea is *denying* it and *movable* is certainly a reasonable opposite to *constant* (*fixed*, according to my dictionary). A similar problem occurs with the supposedly extra word *argue.* The general method for clarifying, explaining, illuminating, or elucidating a proposition in mathematics is to give an argument for it.

The last example indicates the bias of the test maker(2; p. 189):

In each line cross out the word that does not belong there:

Frank, James, John, ~~Sarah,~~ William
~~mechanic,~~ doctor, lawyer, preacher, teacher

After reading these and many other examples, one discovers the not surprising fact that intelligence testers are plagued by the same problem suffered by almost all examiners; namely, it is difficult to ask a question in a general setting that has a precise and unique answer. What may be surprising to one unfamiliar with the methods of mental measurers is the way they resolve this difficulty. An insight into this remarkable resolution is provided by an encounter that Banesh Hoffmann had with the Educational Testing Service. Hoffmann is a physicist and a mathematician who collaborated with Einstein. The ETS, located at Princeton, N.J., is the major source of scholastic examinations in the United States, including the SAT, GRE, and others.

Professor Hoffmann corresponded at some length with ETS(3; pp. 183-97) about the following question that appeared in a booklet, *Science,* published in 1954 by the College Entrance Examination Board.

54. The burning of gasoline in an automobile cylinder involves all of the following except

(A) reduction
(B) decomposition
(C) an exothermic reaction
(D) oxidation
(E) conversion of matter to energy

The desired answer for 54 is E. However, as Hoffmann points out, according to the Theory of Special Relativity, *any* exothermic reaction involves a conversion of matter to energy. This theory was about 50 years old at the time the question was printed. It would be extremely difficult to find a professional physicist or chemist who does not accept it. In fact, I was present at a seminar in relativity theory that was polled by Hoffmann on this very point.

The Educational Testing Service responded to the criticism(3; p. 186): "When such a [superior] student is faced with the above question, he should realize that the classical concepts of matter and chemical change provide the framework in which the question is asked. He also recognizes that the first four processes listed are obviously and immediately involved in the burning of gasoline, and he selects response E as the required answer."

How is the student supposed to realize this when the nonclassical concept of conversion of matter to energy is included in the question?

The reader is referred to Chapters 15, 16, and 17 of Hoffmann's *The Tyranny of Testing*(3). It should be noted that contention and ambiguity can arise even in so "objective" a field as physics. The general attitude of ETS toward this problem will be considered next.

In its pamphlet *Explanation of Multiple-Choice Testing,* the ETS states the following(3; p. 197): "Preliminary tests using the questions that have been prepared can be tried out on students to determine whether, in fact, each individual question discriminates between the better and poorer students or whether there is an element of ambiguity in the question which harms its effectiveness. Such questions can be eliminated or revised in such a way as to avoid ambiguity." In their view, question 54 did not require revision. In other words, the final judgment of a test and the correctness of its answer are based upon the degree to which "better" students tend to perform "better" and "poorer" students tend to perform more "poorly." The process of making this judgment is called *validation.*

In *Intelligence Testing, Methods and Results*(2; pp. 104-12), Pintner describes four criteria against which the tests may be compared. They are: known groups, mainly "feebleminded" and "normal" children; teachers' judgments; school achievement; other tests. All of these depend directly or indirectly upon subjective individual or social appraisals. Nowhere do we see anything resembling precision or objectivity. Intelligence tests are validated by examining their correlation with the fuzzy, personal, idiosyncratic appraisals they are supposed to supplant. In what follows, we shall examine a possible outcome of this system.

Intelligence and Correlation

In view of the key role of correlation in the construction and validation of mental tests, we are led to consider the possibility that correlations obtained in the field between intelligence and such categories as ethnic membership, "success," and ancestry may be somewhat tainted. In fact, I will show that the history and the general pattern of intelligence testing are based upon the following critical factors:

1. A supposedly "standard" group of people whose intellectual abilities are assessed by experts and educators.
2. Experts and educators who, in many cases, are biased by the physical appearance and the ethnic status of the subjects (e.g., skin color, sex, height, facial features, clothing, and mannerisms).
3. The fact that the physical appearance and the ethnic status of the individuals in the standard group affects their lives, their behavior, their reactions to other people, and their reactions to being tested.
4. Questions and preferred answers to a test being chosen so that high scores correlate with the experts' prejudiced view of the standard group.

Now consider a group of people selected so as to show uniform intellectual performance, let us say, by consideration of their employment and competence in that employment. Is it not possible that such a test administered by such experts to such a group will then correlate with the degree of physical and ethnic acceptability of the individuals taking it?

Let us consider these assertions. First, how likely is it that educators and experts involved in the construction of mental tests are biased? To begin with, we must realize that we are considering a group of people who are favored by society. They are not coal miners, assembly-line workers, or sanitation personnel; for that matter, they are not even elementary classroom teachers. They are members of the establishment who have a vested interest in the *status quo*. Moreover, it is very convenient for them to ascribe any inefficiency in the educational system to a lack of innate ability in a sizable proportion of their clients.

The activities and statements of the pioneers of the testing movement around the turn of the century reveal that this is not just a theoretical problem. At that time, a great flux of Italian, Polish, Russian, and Jewish immigrants into the United States threatened to overwhelm the English, Scandinavian, and German stock of this country. To insure the maintenance of "quality," the infant science of mental testing was called to the colors. I quote from Leon Kamin's *The Science and Politics of I.Q.* (4; p. 16):

The first volunteer was Henry Goddard, who in 1912 was invited by the United States Public Health Service to Ellis Island, the immigrant receiving station in New York Harbor. The intrepid Goddard administered the Binet test and supplementary performance tests to representatives of what he called the "great mass of average immigrants." The results were sure to produce grave concern in the minds of thoughtful citizens. The test results established that 83% of the Jews, 80% of the Hungarians, 79% of the Italians, and 87% of the Russians were "feebleminded." By 1917, Goddard was able to report in the Journal of Delinquency *that "the number of aliens deported because of feeble-mindedness . . . increased approximately 350% in 1913 and 570% in 1914. . . . This was due to the untiring efforts of physicians who were inspired by the belief that mental tests could be used for the detection of feeble-minded aliens."*

Imagine being driven from your home by dire political and economic circumstances, selling your belongings to obtain passage in the hold of a ship, enduring extreme discomfort, and risking infection by contagious disease to which some of your shipmates succumb. Now imagine arriving at Ellis Island, a scene of unparalleled confusion, where someone administers an incomprehensible, Alice-in-Wonderland test to you and has you sent back.

The fact that the poor, the homeless, and the victims of prejudice are frequently deprived of physical and emotional order and that many mental testers seem incredibly insensitive to this deprivation is a point to which we shall return later in this article.

Kamin also quotes Dr. Nathaniel Hirsch, a pioneer of the testing movement, as follows(4; p. 28): "'The Jew is disliked primarily because despite physical, economic, and social difference among themselves, "all Jews are Jews," meaning that there is a psycho-biological principle that unites the most dissimilar of types of this strange, paradoxical Natio-Race.'"

It is tempting to quote Kamin's book at much greater length. Instead, I urge the reader to obtain and read this valuable reference. It is worthwhile to quote part of a critique of a paper by Kamin that appears in Loehlin, Lindzey, and Spuhler (Kamin's book is a considerably expanded version of that paper)(1; p. 293):

Kamin's paper has two main parts. The first accuses the pioneers of the U.S. testing movement of more or less deliberately using the intelligence test as a weapon to keep down the underprivileged classes. Kamin finds—as one certainly can—in the psychological literature of the 1910's and 1920's a variety of statements that by today's standards are shockingly bigoted toward ethnic minorities and lower income groups. However, he does not mention the statements that one can also find in that same literature often by the same authors, suggesting that an important motivation of the testing movement has always been to circumvent traditional social class barriers by locating and encouraging talented youth from all sectors of society. The success of this effort is open to some debate—it has probably worked better with the lower middle class than with the really poor. But intent certainly seems to have existed. And many children of impoverished immigrants, among others, have been its beneficiaries.

In short, it is not too difficult to find both good things and bad in the history of the intelligence-testing movement. Kamin, for whatever reasons, has elected to report only the bad.

First, Kamin does quote Binet as opposing the notion of fixed intelligence, saying "'We must react against this brutal pessimism.'"

Second, it should be noted that this critique appears in appendix H. Loehlin, Lindzey, and Spuhler's reference to Kamin's paper in the body of the text consists of a brief dismissal.

Third, consider the phrase, "a variety of statements that by today's standards are shockingly bigoted." The bigots who made these statements were among the educational and intellectual leaders of that age. And, what is more important, in the formative days of the testing movement, bigotry and xenophobia were common and rarely challenged.

Fourth, the assumption that today's standards of racial tolerance are greatly superior to those of the past, while generally held by the white community, is not shared so generally by the black community. Considering the effluences of Jensen, Shockley, Herrnstein, Eysenck, and their apologists, this is not hard to understand.

The "moderates" of the testing movement may argue that they are not responsible for the extremists in their ranks. But these extremists were founders of the movement who set the tone and standards of intelligence testing. The experts of today are not merely writing about past history; they are involved in the construction and administration of tests that affect the lives of individuals as well as social policy. They should be deeply concerned about the existence of corrupting influences that affect the construction of the tests and their so-called validation. There is little evidence of this concern beyond the "tsk, tsk" level (see, for example, p. 232 of Loehlin, Lindzey, and Spuhler). Instead, they seem to be more annoyed at people like Kamin for calling attention to the problem.

Turning to my third assertion, it should be obvious that one's ethnic background influences one's experiences and one's behavior. What is more, there should be little opposition to the notion that major physical characteristics like sex and skin color are also influential. But we should not forget that less socially significant characteristics have an important effect on experience and perception as well. For example, height, weight, wearing glasses, set of eyes, sizes of ears, and so forth, all fit into popular stereotypes about personality and thus affect, in a nontrivial manner, an individual's social environment. For example, a short person has one view of the world and may logically answer questions one way, while a tall person may logically answer them another way if the questions implicitly or explicitly involve a concept related to the value of height.

Let us now consider my last assertion, that an "intelligence test" could be designed, consciously or unconsciously, to rate the subjects according to acceptability.

In the case of a verbal exam, it is plausible that we can design it so that the results correlate with an arbitrarily chosen status scale by writing the questions wholly or partly in the dialect of the "high status" subjects and choosing answers that agree with the logic and life experience of these "superior" individuals. On the other hand, it might not be as easy to see how we could do this with a nonverbal exam. For example, Raven's Progressive Matrices seem at first glance to be unconnected with anything but abstract problem-solving ability. But the problems in this jigsaw-puzzle-like exam depend on patterns and order. The subject must complete a two-dimensional pattern, choosing a piece from a given set, in such a way as to satisfy a certain principle of order and neatness. It is probable that the degree of order and neatness in one's life might have a great deal to do with how one might perform on this exam. Furthermore, patterns are certainly not culture free. Architecture, home furnishings, and clothing all condition us to desirable and undesirable patterns, to mention just three cultural influences on our pattern perception.

It would be interesting to construct a counter exam, Schwartz's Regressive Matrices. In my exam, the subject is asked to complete a pattern that is a little disorderly to begin with. He or she must choose from a set of pieces, none of which is entirely correct. The subject, in other words, must make do in a disorderly situation. Moreover, the exam should not be held in the middle-class environment of an examiner's office or a classroom. The standard environment for the exam should be a tenement building with rats scurrying behind walls, cockroaches running around in the corners, and someone being mugged outside on the street.

Now, it is not even necessary to consciously and openly design the exam to favor a certain group. If that group is designated as superior by some establishment, then all we have to do is monkey around with the questions and answers until we get something that correlates with this status assignment. If we already share the value system that assigns status, then we may very well build in that correlation to begin with and use the observed correlation as "validation." We have seen evidence indicating that this, in fact, was done.

Sophisticated proponents of IQ tests argue that there is nothing wrong with this: that IQ scores do correlate highly with "success," that is, high income and social status. (Note that they do not assert that IQ scores correlate with job competence[5; p. 16].) Furthermore, they say, IQ is a predictor of ability to get along within the "system." This may be so, if getting along means conforming. The utter weakness of this defense is that the world is rapidly and radically changing socially, politically, and economically. We cannot predict which skills will be valuable, which will be obsolete, and which will be antisocial in the future. One thing seems clear, the "successful" people have brought this country (consider New York City) and several others to the brink of ruin. The "system" is rotten.

Flexibility and novelty of thought, inventiveness, and creativity, as well as the ability to deal with disorderly and confusing situations, are not rated highly by IQ tests. Indeed, they may lead to low scores. Thus, there is good reason to suspect that IQ testing may discourage and suppress many who might otherwise make needed contributions to society.

THE STATISTICS

This jaundiced view of the social scientific use of intelligence testing may seem to fly in the face of nearly three-quarters of a century of data collection by dedicated scientists. In particular, the studies on separated identical twins are often quoted as very strong and direct support for the hereditarian position. These studies suggest, it is asserted, that, if identical (monozygotic) twins are separated at birth and raised in uncorrelated environments, then their IQs will still be highly correlated. This is powerful medicine for it supports not only the hereditarian view of intelligence but also the ability of the tests to measure some innate ability.

I submit that the basic theoretical foundation underlying the statistical analysis of the data is unacceptable, that the conditions of data collection were, in many cases, nothing like the ideal situation described above, that their interpretation is open to question, and that some of the reports would suggest to all but the faithful a strong hint of hanky-panky.

The Founding Fathers

The introduction of the application of statistical theory to the study of human affairs is generally attributed to Lambert Quetelet. Quetelet was director of the Royal Belgian Observatory and a tutor of Prince Albert. He constructed a statistical picture of humanity inspired by the theory of astronomical observations that had been developed by Laplace, Legendre, and Gauss. According to this theory, when we make a series of observations of the position of a star, the object of our observation has a true and definite value; however, the values that we see and record are actually that value *plus* a series of errors resulting from a very large number of small, random disturbances. The theory tells us that these errors have a certain statistical pattern; that is, they are *normally distributed.* The French mathematician and physicist Henri Poincare once remarked that physicists generally assume that mathematicians have proven the validity of the "Normal Law" by means of mathematical reasoning, while mathematicians assume that physicists have substantiated it with physical arguments. Nevertheless, the use of the Normal model in astronomy and physics is pretty safe.

The translation of this model to the study of human behavior is another story. It suggests that the human population corresponds to a series of randomized observations of a standard ideal individual, Quetelet's "Mean Man." Observed variations are a result of hereditary differences and randomness in the reproductive process that are not subject to complete understanding, let alone remedy.

The ideas of Quetelet were not well received until Sir Francis Galton adopted and promoted them. Galton was a founder of the eugenics movement. His first book, *Hereditary Genius*(6; p. 1), begins with the sentence, "I propose to show in this book that a man's natural abilities are derived by inheritance, under exactly the same limitations as are the form and physical features of the whole organic world." In addition to his own writing, Galton contributed to the development of the statistical study of inheritance by endowing a research laboratory headed by Karl Pearson. Galton and Pearson rejected the application of Mendelism to the study of human inheritance, preferring the so-called Law of Ancestral Inheritance. As a result, when William Bateson tried to publish findings that supported Mendelism, Pearson refused to accept them in his journal, *Biometrika.*

The last and most recent contributor to the field of biometry I shall mention here is R. A. Fisher, an admirer of Galton. Fisher's *Statistical Methods for Research Workers*(7) is widely used by natural and social scientists. The copy I have at hand is from the eleventh edition. A considerable part of this book is devoted to promoting statistical tests of significance and other mathematical recipes for determining whether or not to accept at face value the apparent results of an investigation. For example, on pp. 94-95, Fisher considers a study of 13 monozygotic twins who were criminals and 17 dizygotic twins who were similarly wayward. By examining the frequency of criminality among their siblings (heavy for the monozygotic, light for the dizygotic), Fisher concludes, by application of the chi-square criterion, that criminality is *significantly* more frequent among monozygotic twins of criminals than among dizygotic twins of criminals and that the former are more alike in their social reactions as a result of genetic factors. Thus, a mechanically applied statistical test and a questionable quantification of statistical significance are supposed to convince us that criminality is inherited.

It seems incredible that so complex and serious an issue can be resolved "significantly" by a study of 30 twin pairs and a mindless arithmetic calculation. Yet, Fisher's book is in its eleventh edition (at least). Unfortunately, social scientists are not likely to have the time or the inclination to look for the derivations and the precise statements of the conditions of applicability of Fisher's formulae. These derivations and statements do not appear in the book. It is easy enough to be impressed by a wealth of mathematical symbols, not to mention the list of 10 academic degrees and titles that accompany Fisher's name on the title page.

It is ironic that this arch-proponent of the unquestioning application of statistical recipes should become very circumspect about using statistical analysis to determine the relationship of smoking to lung cancer. Here Fisher is unwilling to accept the apparent implication of the statistics at face value. He argues that the correlation between smoking and lung cancer is not due to a causative relationship but rather a result of a common genetic propensity to smoke and to contract cancer. He supports this contention by citing studies of twins. (It is my conjecture that the observed correlation of smoking and twinship is a result of their becoming very nervous from being constantly bombarded with studies by statisticians and intelligence testers.) The reader may be interested to learn that, while professing this interesting theory, Fisher was employed as consultant to the Tobacco Manufacturers' Standing Committee(8).

It is not my intention to argue against scientific observation or the statistical analysis of data collected by such observation. I do wish to discourage the mindless application of cookbook formulae put forth by "authorities" who, I claim, have a definite aristocratic bent and an ideological axe to grind. Fisher's book encourages the use of *ad hoc* methods that are very handy for justifying the hereditarian position. One can also use *ad hoc* mathematical and statistical arguments in opposing the hereditarian position, but the historical conditions of the development of genetics and psychometry have tended to promote the use of "pragmatic" mathematics in support rather than opposition.

The Statistical Model of the Inheritability of Intelligence

Now consider how this questionable quantity known as IQ is supposed to be distributed in the population and how this distribution is measured. Let us, for the sake of argument, assume that there is a measurable thing called *general intelligence* and see how we are to deal with its observation in the spirit of the Quetelet-Galton-Pearson-Fisher approach. We shall assume that there is a mean value of intelligence (designated by M) that is constant. From this mean, we assume there are "random" variations and that each individual variation is the sum of two components: one component being the variation due to hereditary differences (designated by H) and the other component being the result of environmental differences (designated by E). Thus, the intelligence of any given individual (designated by I) can be written as the following sum:

$$I = M + H + E$$

A word used frequently in this business is *heritability*. Roughly speaking, heritability is calculated by dividing the spread or dispersion of values of H in the population by the spread of values of I in the population. Heritability of 100 percent would indicate that differences in intelligence are due entirely to genetic causes. Heritability near 100 percent would indicate, according to this theory, that worrying about environmental differences is mostly a waste of time, since they have little effect.

Let us consider this model carefully. First of all, there is no way one could measure the hereditary component alone or the environmental component alone. Every human being has a hereditary background and is bathed in an environment at all times. There is no conceivable way to screen either factor out. The exercise of intelligence, if it is anything, is the use of inherited faculties that have been modified by the environment to deal with and respond to situations arising in the environment.

Second, even if we believe that two such separate components exist in some Platonic sense in that great psychology department in the sky, their combination might not be additive; it might, for example, be multiplicative. Given the interactive nature of intelligence, a multiplicative model would seem more plausible.

Unfortunately, introducing an interaction term makes statistical analysis difficult, since the mathematical form and probability distribution of this term are unknown and either might be very complicated.

R. J. Light and P. V. Smith have studied the possible effects of interaction by means of a Monte Carlo computer-simulation technique(9). By adding a very small interaction term (1 percent of the total variance) to their model, they cause a decrease of the mean IQ from 91.26 to 86.81 in their hypothetical black population but no decrease of the mean in the white population. An interaction term of 10 percent further depressed the mean of the hypothetical black population to 82.59 but left the white mean unchanged. It should be noted that Light and Smith had already explained the subnormal mean of 91.26 as arising from an adverse environmental distribution in their first model, which was additive (noninteractive).

They also showed that estimating the interaction variance by computing correlation coefficients (as done by Jensen [5; p. 39 ff.]) was subject to a standard error of 9 percent even in a situation unrealistically favorable to this approach. Thus, an interaction variance of 10 percent might well appear as only 1 percent. Thus, the work of Light and Smith indicates that an unnoticed interaction component may invalidate our statistical computations.

In a later book, Jensen presents another method for assessing the amount of interaction present(10; p. 322 ff.). Citing Jinks and Fulker(11), Jensen computes the correlation between the mean IQ of each twin pair and the absolute difference between twins. This correlation turns out to be -0.15, which he asserts is not "significantly different from zero." He concludes: "These data, then, do not show evidence of a genotype × environment interaction for IQ."

This last sentence is very slippery. It is quite possible that Jensen saw no such evidence; however, the correlation he calculates does not eliminate, by any means, this possibility. One can construct an analysis of a variance model in which over 80 percent of the variance is due to interaction, and yet in this model the correlation cited is exactly zero (see appendix, pp. 35-36).

The point is, even if this correlation vanishes when interaction is absent, it is a fundamental error to assume that, because it vanishes (or seems to), there is no interaction. This is not the only error in basic logic made by Jensen and company, as we shall see.

Another complicating factor that makes the definition and computation of heritability difficult is the statistical covariance of genotypes and environments. This is a tendency for certain genotypes to be more likely to be found in certain environments. Notice that interaction deals with the way genes and environments act while covariance deals with the way genes and environments are found. Given family, ethnic, and racial residential and employment patterns, the magnitude of covariance must be considerable.

Let us see how Jensen deals with this problem(5; p. 39):

To the degree that the individual's genetic propensities cause him to fashion his own environment, given the opportunity, the covariance (or some part of it) can be justifiably regarded as part of the total heritability of the trait. But if one wishes to estimate what the heritability of the trait would be under artificial conditions in which there is absolutely no freedom of variation in individuals' utilization of their environment, then the covariance term should be included on the side of environment. Since most estimates of the heritability of intelligence are intended to reflect the existing state of affairs, they usually include the covariance in the proportion of variance due to heredity.

According to this approach, a black's genetic propensities cause him or her to fashion his or her birthplace as a ghetto. The children of immigrants cause them to fashion parents with heavy accents, alien customs, and so forth.

Let us consider one more example of Jensen's logic(5; p. 238), in which he replies to a paper by J. McV. Hunt: "What Hunt is saying, essentially, is that the correlation between IQ and occupation (or SES) is due entirely to the environmental component of IQ variance. In other words, this hypothesis requires that the correlation between genotypes and SES be zero."

To show that the statement that he derives is false, Jensen introduces a fancy formula,

$$r_{12}^2 + r_{13}^2 + r_{23}^2 - 2r_{12}r_{13}r_{23} < 1$$

that is nearly correct (the strict inequality should be replaced by a weak inequality). But this is irrelevant, since Jensen's "in other words" statement does not follow from what he claims Hunt is saying. A genotype (for example, skin color) might determine its bearer's occupation. The occupation, in turn, might determine the measured IQ. We could then find 100 percent *correlation* between genotype, SES, and IQ, despite the fact that the *determination* of IQ performance was entirely due to occupation and not dependent on genotype.

Throughout Jensen's article, we see statistical formulae that are dubious, irrelevant, and misapplied, together with erroneous logic that is used to obscure the plain facts of social causation.

The Data

Up to this point, I have mentioned a concept of intelligence that is simplistic, a method of measuring it that is unimpressive, and statistical methodology that is full of unreasonable assumptions. Nevertheless, on the basis of this theoretical edifice, a plethora of data has been collected and analyzed. If the data and calculations were very consistent and showed strong correlations between IQ and genotype, it might appear that the foregoing was simply wrong-headed, metaphysical speculation. We must, therefore, consider these data.

The strongest, clearest, and most direct support for the heritability of intelligence allegedly comes from the study of separated monozygotic twins. In all other studies, there is a strong possibility of the interaction and covariance of the hereditary and environmental contributions to intelligence. But, in the case of monozygotic twins, the genotype is identical and the only variation in a given twin pair is environmental. A great deal of effort has been expended in measuring the variation of IQs within twin pairs and comparing it with the variation between twin pairs.

A statistical calculation yields a number called the *intra-class correlation coefficient,* which purports to indicate the sameness of members of any MZ pair. An intra-class correlation coefficient near 1.0 for twins who have been reared in separate households with uncorrelated and widely varying environments would be evidence in favor of the hereditarian position that their IQ is determined mainly by the genotype and has little to do with the variation of the environments involved in the study.

As I have indicated, the credibility of the statistical procedure is open to question. But, for the time being, I will follow the critique by Kamin, who apparently accepts the reliability of conventional statistical procedures and modern IQ tests. The question is: do separated twin studies really reveal high correlations in IQ between MZ twins reared in truly uncorrelated environments? Jensen states that there have been only three major studies of MZ twins separated early in life and reared apart. Of these, he says Sir Cyril Burt's study is the most interesting(5; pp. 51-52). Burt's work involves more twin studies than anyone else's. He obtains higher correlations than anyone else. And he alone provides quantitative data and alleges that the environments in which the separated twins were reared were not correlated. Note that in this regard Burt is not only quantifying intelligence but also the quality of a child's environment.[1]

The Burt papers contain a wide array of questionable items(4; pp. 35-47), including:

1. In a 1943 kinship study, there is no indication of how the subjects were selected or how their IQs were measured. There is no presentation of the raw data, the majority of which "remain buried in typed memoranda or degree theses."
2. There is no indication of the distribution of the children among various economic categories.
3. The 1943 paper reports a correlation between IQ and *economic* status of .32. In a 1956 discussion of the same study, we are told that the ratings of the "*socioeconomic*" status and the "adjusted assessments" of intelligence were .315, which rounded off agrees with the earlier .32. However, *economic* has changed to *socioeconomic,* despite the fact that Burt had earlier made a point of distinguishing his strictly economic status from cultural status for which other researchers had obtained higher correlations. Moreover, we are informed in the 1956 paper that the correlation between "crude test results" and socioeconomic status is .453. Thus, the earlier reported results were based on some sort of "adjusted assessment," and this is revealed 13 years after the initial report.
4. In many of Burt's papers, the correlation coefficients are given to three decimal places. This gives the unwary reader an impression of precision that cannot possibly exist in a study such as this.
5. In the 1955 and 1966 studies of siblings and dizygotic twins, Burt obtains correlation coefficients, again to three decimal places, that show fantastic stability. That is, measurements of different groups of different sizes at different times turn up measured values that are equal to two or three decimal places.
6. Burt, in 1958, refers to his method of "adjusted assessments" as follows: "The final assessments for the children were obtained by submitting the marks from the group tests to the judgment of teachers . . . where the teacher disagreed with the verdict of the marks, the child was interviewed personally, and subjected to further tests, often on several successive occasions"(4; p. 39).
7. In 1957, Burt said that the raw IQ is far less trustworthy than the judgment of a child's teacher. In 1943, he said the precise opposite.
8. There are a number of anomolous differences between the variances within subgroups that, according to the theory, should not exist.

Richard Lewontin calls Burt's work the Watergate of genetics. Now we come to the cover up. We again quote Loehlin, Lindzey, and Spuhler(1; p. 294):

Kamin's scrutiny of Burt's published work amply demonstrates what is in fact clearly the case—that Burt's empirical studies in this area are inadequately and often carelessly reported, at least in sources readily available to the U.S. investigator. If one wishes to use Burt's data, one must take a lot on faith—which Kamin clearly is not inclined to do. Alternatively, one could presumably attempt to find out *the explanations for some of the anomalies in the data: while Burt himself is dead, doubtless some*

[1]A recent article in the *London Sunday Times* (October 24, 1976, pp. 1-2, by Oliver Gillie) reports that Sir Cyril Burt, whose studies are used to support the hypothesis that intelligence is largely inherited, probably faked much of his data. Apparently, Burt "often guessed at the intelligence of parents he interviewed but later treated these guesses as hard scientific data," named collaborators who may never have existed, and manipulated his data to fit his hypotheses. Burt's ideas were influential in moulding British education. Obviously, these new findings are of the utmost importance.

of his former students and research associates could shed light on the details of some of the research, and it might not be out of the question to track down some primary documentation of the studies. Kamin prefers simply to dismiss Burt's data as "not worthy of serious scientific attention."

It is hard to see how they can characterize a 13-page critique as a simple dismissal. What is more remarkable is the implication that it is the responsibility of the faithless Kamin to clean up Burt's mess.

Kamin also deals with a variety of other separated MZ twin studies. Again, he finds a number of serious flaws. It is instructive to go directly to the response of Loehlin, Lindzey, and Spuhler(1; p. 294):

Next, Kamin turns his critical eye on the three studies of identical twins reared apart other than Burt's, those by Newman, Freeman, and Holzinger (1937), Shields (1962) and Juel-Nielsen (1965). He correctly notes that the degree of separation in such studies is often exaggerated in secondary accounts: in Shield's study, for example, the criterion for inclusion was merely having been reared in separate homes for at least five years during childhood, and in many cases the homes were those of relatives. Still, even those homes are presumably less alike than the same *home, and so it is not at all clear, on Kamin's hypothesis, why separated identical twins in all these studies should be more highly correlated in IQ than typical like-sex fraternal twins who are reared together.*

This is a serious point that Kamin does, in fact, answer, but his answer is perhaps not emphatic enough. As a rule, identical twins are identified as such by their appearance (and sometimes by their behavior). That is, in most cases, identical twins are phenotypically identical and the identity of their genotype is inferred, not certain. Without this inference, the correlations we observe are only between phenotypes. It is, as I have argued before, entirely possible that appearance phenotype is correlated with "intelligence" phenotype. A child's height, weight, vision, speech patterns, posture, forehead size, and so on, all affect the attitude of his or her parents, teachers, friends, and intelligence testers. Recall the subjective "reassessment" in Burt's measurements. Jensen also mentions routinely retesting children who "appear" brighter than their test scores(5; p. 100).

Spreading the Word

The final point I will discuss is the dissemination of information collected by researchers. There is increasing concern among some social scientists over the difficulty of establishing through empirical studies that differences between groups and individuals do *not* exist. This, of course, is very troublesome to an egalitarian. There are, to begin with, three more or less technical or at least accidental reasons:

1. Zero, as mathematicians of my stripe would say, is a very unstable or nongeneric number. If you are measuring a difference that in fact is zero, any slight perturbation or error will mask it and the difference will be measured as a nonzero value. On the other hand, a nonzero number has some stability, for it is surrounded by its nonzero comrades, so to speak; thus, its measured value will also be nonzero. This instability of zero leads to great difficulty in using statistical methods to find the little bugger.

2. A result establishing or supporting the nonexistence of a difference tends to be less dramatic, attracts less attention, and is harder to get published than results supporting the existence of a difference. There is, therefore, a natural tendency for the literature to be skewed against the hypothesis of equality.

3. There is another more important problem: editors and researchers themselves may screen out results and articles that tend to deny the existence of sex, race, and ethnic differences in various traits for plain-old political reasons.

If this last point strikes you as paranoid, don't forget Karl Pearson, the author of *The Grammar of Science,* who excluded William Bateson's work from the pages of *Biometrika* because it ran counter to his own pet theory of heredity.

CONCLUSION

In summary, I have asserted that:

1.. The psychometric concept of intelligence is highly oversimplified.

2. IQ tests have no *a priori* incisiveness, they are boring, and they probably reflect the social prejudices of the constructors and administrators.

3. Many of the pioneers of the testing movement are on record as bigots and xenophobes.

4. These built-in prejudices, perpetuated by a questionable validating procedure, might alone account for observed correlations with ethnic categories.

5. Although the test scores may correlate with success (i.e., socioeconomic status achieved), they do not seem to correlate with competence in employment and may very well correlate negatively with the desirable skills of inventiveness and the ability to function well in difficult environments.

6. A theory of statistics pioneered and developed by hereditarian fatalists is used by many people who seem not to understand its derivation or limitations. This is the theory used to substantiate the hereditarian position.

7. The methodology of collecting and analyzing data in frequently cited twin and other IQ studies is riddled with flaws.

8. The reporting of research is quite possibly biased in favor of the hereditarian position.

The "science" of intelligence testing was invented and developed to serve the requirements of the social class to which the inventors and developers belonged. Findings of such a science should be subjected to the severest scrutiny. Instead, the leaders of the intelligence testing movement have shown shocking prejudice and laxity. There is no reason to continue calling intelligence testing a science. Like the divine healing power of the Stuarts, it is simply peddling hokum to justify privilege.

References

1. Loehlin, J. C., Lindzey, G., and Spuhler, J. N. 1975. *Race differences and intelligence.* San Francisco: W. H. Freeman and Company.
2. Pinter, R. 1931. *Intelligence testing, methods and results.* 2nd ed. New York: Henry Holt and Company.
3. Hoffmann, B. 1962. *The tyranny of testing.* New York: Crowell-Collier Press.
4. Kamin. L. 1974. *The science and politics of I.Q.* New York: Halsted Press.
5. Jensen, A. R. 1969. How much can we boost IQ and scholastic achievement? *Harvard Educational Review* 39:1-123.
6. Galton, F. G. 1869. *Hereditary genius: An inquiry into its laws and consequences.* New York: Macmillan.
7. Fisher, R.A. 1950. *Statistical methods for research workers.* 11th rev. ed. London: Oliver and Boyd.
8. *British Medical Journal.* 1967. Vol. II. (Correspondence by R. A. Fisher [p. 43], R. N. C. McCurdy [p. 158], and R. A. Fisher [p. 297].)
9. Light, R. J., and Smith, P. V. 1969. Social allocation models of intelligence. In *Science, heritability, and IQ.* Reprint series no. 4. *Harvard Educational Review.*
10. Jensen, A. R. 1973. *Genetics and education.* New York: Harper and Row.
11. Jinks, J. L., and Fulker, D. W. 1970. Comparison of the biometrical, genetical, MAVA, and classical approaches to the analysis of human behavior. *Psychological Bulletin* 73:311-49.

APPENDIX

We shall exhibit a hypothetical example in which the correlation between pair means and absolute pair differences is zero, which, at the same time, contains marked (82 percent) interaction variance. This shows that a zero value of this correlation does not imply the absence of interaction.

Consider a population with equal numbers of twin pairs of types 1 through 4. In each pair, the score of twin 1 is listed in column 1 and the score of twin 2 is listed in column 2, as follows:

	Twin 1	*Twin 2*
Type 1	130	90
Type 2	70	150
Type 3	110	70
Type 4	130	50

The score for twin j of type i, S_{ij}, may be written as the sum of the population average, M, the pair type effect, $m_{i.}$, the intra-pair effect, $m_{.j}$, and the interaction effect, $R(i,j)$. Thus,

$$S_{ij} = M + m_{i.} + m_{.j} + R(i,j)$$

We note first that the population average is 100 and add the marginal averages to the table. The averages are written as the sum of M plus the row or column effects.

	Twin 1	*Twin 2*	
Type 1	130	90	$100 + 10 = M + m_{1.}$
Type 2	70	150	$100 + 10 = M + m_{2.}$
Type 3	110	70	$100 - 10 = M + m_{3.}$
Type 4	130	50	$100 - 10 = M + m_{4.}$
	$100 + 10 = M + m_{.1}$	$100 - 10 = M + m_{.2}$	

We now write each term, S_{ij}, in the table as the sum of $M + m_{i.} + m_{.j} + R(i,j)$, underscoring the interaction, $R(i,j)$.

	Twin 1	*Twin 2*
Type 1	$100 + 10 + 10 + \underline{10}$	$100 + 10 - 10 - \underline{10}$
Type 2	$100 + 10 + 10 - \underline{50}$	$100 + 10 - 10 + \underline{50}$
Type 3	$100 - 10 + 10 + \underline{10}$	$100 - 10 - 10 - \underline{10}$
Type 4	$100 - 10 + 10 + \underline{30}$	$100 - 10 - 10 - \underline{30}$

We compute pair mean/absolute difference covariance:

$$\Gamma = \tfrac{1}{4} \sum_{i=1}^{4} \left(\frac{S_{i1} + S_{i2}}{2} - M\right)\left(|S_{i1} - S_{i2}| - 60\right)$$

$$= \Big(10\cdot(-20) + 10\cdot(20) + (-10)(-20) + (-10)(20)\Big)/4$$
$$= 0$$

It follows that the correlation is zero.

However, the interaction variance, V_I, is

$$\tfrac{1}{8} \sum_{i,j} (R_{ij} - 0)^2 = 900$$

while the total variance, V, is

$$\tfrac{1}{8} \sum_{i,j} (S_{ij} - M)^2 = 1100$$

Thus,

$$V_I / V \doteq .82$$

IQ and Scientific Racism

Val Woodward

Science, like health care, education, and welfare, is an institution created and maintained by society, and, like other social institutions, it influences and is influenced by society. Ideology shields this reality to the extent that, in the popularized view, first, science is *autonomous* of other societal concerns (e.g., ethics, politics, history); second, science is indicative (i.e., science can tell us *how* to do what we want to do, but not *what* we should want); and, third, the "objectivity" of science insures its social and political *neutrality*. By ignoring the interrelationships of science and society, such reasoning neatly sidesteps the causes and consequences of scientific racism simply by ignoring its existence. However, as one analyzes the relationships between science and society, it becomes increasingly apparent that science, by its very nature, reflects and shapes the characteristics of the society of which it is a part.

One dominant characteristic of our society is racism. A related characteristic is the existence of a divided, diversified, and "inexpensive" labor force. In addition to these objective social relationships, there are rationalizations designed to explain the disparity between rich and poor and among segments of the labor force. It is the thesis of this paper that racism has rubbed off onto science and that science is an accomplice to racism to the extent that it provides rationalizations for the dominant class to explain its control over and exploitation of the working classes.

Although no single definition of racism is satisfactory in every aspect, racism, as used here, is the institutionalization of inequality, rationalized by real and imagined differences among groups of people (a real difference is the fact that non-whites earn less money than whites, and an imagined difference is that non-whites are less deserving than whites). Within this context, racism is not prejudice or bigotry, attitudes that many individuals hold toward ethnic and racial minorities. Institutionalized inequality means differential access to good jobs, education, health care, legal services, and so on. Differential access to jobs and social services is a driving force that both maintains institutional racism and fosters bigotry and prejudice.

The reality of institutional racism is demonstrated by the disproportionately low numbers of persons within each minority group who become doctors, teachers, owners, and managers, as contrasted with the disproportionately high numbers who are miseducated, unemployed, imprisoned, and economically exploited. However, the objective existence of institutional racism does not, by itself, inform us of the social and political mechanisms that keep it alive and well.

Economic necessity is basic to the maintenance of racism. Shortly after Columbus landed in the West Indies, the economic potential of sugar farming began to turn real profits, and from this came the slave trade. On the heels of slavery came the rationalizations justifying it (better a Christian slave than a free heathen, etc.). Religion was important at first, but soon thereafter science began to compete with religion as the arbiter of natural phenomena. Scientific theory began to provide rationalizations for economic and class disparities among groups of people.

Today, we hear from racist biological determinists that there are fewer native American chemists because there are fewer native Americans with the intelligence to become chemists. Biological determinists take the position that the structure of native American societies (which they imply is genetically determined) fails to provide the incentives and preparation for disciplined and rigorous mental work. And, of course, we still hear the argument that there are fewer native Americans who want to become doctors. There are many variations on these racist themes.

An aspect of these so-called scientific explanations of racism that can be easily overlooked is that a theory itself can be racist. In other words, an explanation of race differences can have the power of a self-fulfilling prophecy. Consider the following: scholarly observers declared after examination of antebellum Southern plantations that whites must be more intelligent than blacks—after all, some whites were sipping mint julep while the blacks were doing the dirty work. The theory was developed that whites are smarter than blacks, and the evidence was provided by the fact that their respective social roles were different.

Even after science had set upon developing additional measures of intelligence, there was only one referent for intelligence available, and that was success as determined by social role. Therefore, all tests that have not correlated with the meritocracy have been discarded, and extant tests have conspired to "prove" that blacks are less intelligent than whites. Consider the self-fulfilling prophetic quality of such a test. A measure of intelligence exists; it is standardized by success and therefore it correlates with success. The test results are used to *track* the children into one or another level of the meritocracy, and, following this, the low test scores of the Teamsters and the high test scores of the president's advisors are said to correlate with success/failure and are determined by genes.

In analyzing the activities of science, I do not mean to imply that all scientists are plotting and scheming to rationalize social structure by theory. But, because there is a mystique (ideology) about science, there is a built-in tendency to believe that science is free of human fallability or that only applied science, not pure science, is tainted by human imperfection. Scientists claim objectivity, adhering to the philosophy of logical positivism (data are our only access to reality). But, if one examines science as object, the contradictions between the mystique of science and the social-political functions become apparent(1). The contradictions have to do with the function of ideology, which is to interpret reality—most often by shielding the relationship between the proponents of the ideology and the material power base of society. This is illustrated by the following example.

After giving the matter some thought, Aristotle proclaimed that a big rock will fall faster than a small rock. We might wonder why he didn't climb a tree and drop some rocks, just to check it out. But Aristotle didn't share our biases about the value of data; to him, thought was superior to experience, meaning that an experiment would only have clouded the issue. Aristotle proclaimed that the sun revolves around the earth, and on and on. What is important to the analysis of science in society is not that Aristotle believed thought to be superior to experience but that an ideological paradigm was erected around this belief such that its critical examination was effectively forbidden for more than 2,000 years!

Two millennia later, Copernicus, Kepler, Galileo, and others began to question the clear thinking of Aristotle and to approach reality by way of experience. As you recall, these activities got them into a peck of trouble with the authorities—the guardians of ideology—not because it mattered whether little rocks fall as fast as big rocks, but because of the threat to their ideology posed by another access to reality. The authorities had no intention of giving up their power, and they forbade mutation of any of the elements they believed necessary to keep it intact. In sum, there existed at the time a tight-knit and dominant ideology, which, by agreement of those in power, had become internally self-consistent. This ideology was integral to the economic, political, and class structure of the time.

It is commonplace today to accept experience as being an important adjunct to the construction of models of reality; it is equally as commonplace to assume that experience forfends ignorance as well as leading us toward reality. The comfort provided by this knowledge and this assumption can make us vulnerable both to the weaknesses and the misuses of empiricism. For example, we can become susceptible to the notion that science, the generic name given to the kinds of experience that lead to better understandings of reality, is free of ideological taint. Historically, science is said to be autonomous of ethics, religion, politics, and economics and neutral with respect to value judgments (data are value free). The assertions that science is autonomous and neutral are patently false, as, for example, even a cursory examination of the eugenics movement shows.

Not only is science shaped by the society housing it, science profoundly influences the character of society—attitudinally, politically, and ethically, as well as physically. The dominant ideology can shield us from these relationships, especially those that arise from the acquired characteristics of science. For example, there has been the belief in science that our only access to reality is by way of data (logical positivism); that all phenomena, cultural, behavioral, and physical, are explicable by natural law (reductionist logic); and that the abstract quality of theory justifies the separation of "pure" from applied science (experience guided by theory is objective, while experience guided by need, or profit, is subjective).

That the character of science is influenced by society is an objective reality. To the extent that this happens, science can subjectively lend "objective" support to the dominant ideology. We have witnessed the complicity of science in war, exploitation of Third-World peoples and prisoners for experimental purposes, support of industries that produce insecticides, herbicides, aerosol sprays, and food additives, and so on; in this paper, we illustrate the complicity of science with policy makers who base social policy upon theories of human nature.

THE SCIENCE OF HUMAN NATURE

There have been many explanations of human nature during the past three or four thousand years. This history has seen the mergence of curious admixtures of science and religion, and even today religiosity within human-nature theories is prevalent. The two theories that have dominated all others and that are today in vicious competition as arbiters of social polity are the hereditarian and environmentalist theories of human nature.

Environmentalists have taken the position that the environment is the major determinant of human nature and that, therefore, the equalization of human environments and opportunities will lead to greater equality of achievement among individuals, groups, and races. Environmentalists often have worked for egalitarian reforms in education, health care, legal services, jobs, and so on. Hereditarians, on the other hand, argue that human nature is determined primarily by genes. Therefore, they take the position that human nature is inflexible, or intractable, and that social policy should be adjusted to serve the inherited capacities and differences among individuals, groups, and races. Modern hereditarians argue, for example, that persons carrying genes for low intelligence cannot profit from a superior education and that to provide people of low intelligence with a superior education is a waste of scarce resources.

The most recent resurgence of the hereditarian theory of human nature has been advanced by Arthur Jensen, an educational psychologist at Berkeley. In an article entitled "How Much Can We Boost IQ and Scholastic Achievement?"(2) Jensen answers, not much; he asserts that compensatory education has failed and that the reason it has failed is because the genes of certain groups of people (particularly blacks) will not allow them to profit by it. It was the main thesis of his 123-page paper that "scientific evidence" tells us that blacks are less intelligent than whites.

In the September 1971 issue of *The Atlantic Monthly,* in an article entitled, "I.Q."(3), Richard Herrnstein, a behavioral psychologist at Harvard, transplanted Jensen's thesis to the class structure of society. Herrnstein says that the entire class structure of society (that is, the meritocracy) is determined by genes (except for a "few sweepstakes winners and starving geniuses").

Shortly after Jensen's paper was published, William Shockley(4), Nobel Prize winner (coinventor of the transistor) and physicist at Stanford University, drew up a legislative proposal calling for sterilization of persons with IQs below 100. He proposed to begin the program with persons on welfare and to entice them with the promise of $1,000 per IQ point below 100 to "volunteer." His proposal was made palatable to conservative legislators by arithmetic used to show that the program would pay for itself in two or three generations by eliminating the need for welfare programs, prisons, and mental institutions. At about the same time, Daniel Moynihan, then advisor to President Richard Nixon, announced that "'the winds of Jensen' were gusting through the capital at gale force" (quoted by Neary [5; p. 58D]).

Both the content and the racism of this so-called scientific assertion have a long history. Only a summary can be presented here, but such a summary makes it possible to compare some of the parts of the IQ argument with the historical setting within which the parts took form.

Norman Daniels(6) summarizes the IQ argument as follows:

1. IQ measures intelligence, a central behavioral trait.
2. IQ is an important causal determinant of success in school and in life.
3. IQ is highly heritable (80 percent) in the white population.
4. Mean differences in IQ between black and white populations (15 points) and between social classes are large.
5. Therefore:
   (a) IQ is resistant to change.
   (b) Inequalities in IQ cannot be eliminated.
   (c) Black/white and social class IQ differences are probably genetic in origin.
   (d) Attempts to achieve greater equality in a variety of social contexts (e.g., education) are unrealistic.

(For discussions and criticisms of premise 1, see Daniels(6) and Block and Dworkin(29); premise 2, see Bowles and Gintis(7); premise 3, Kamin(8) and Feldman and Lewontin(9); and premise 5c, Lewontin(10). If premises 1-3 are discredited, then premise 4 is trivial and premise 5 is wrong.)

The history of premises 1 and 2 shares much in common with the story of Aristotle and the rocks. Intelligence is defined by authority, and its predictive value as a determinant of success arises from the fact that intelligence tests are normalized against "success." Premise 3, on the other hand, is a collection of numbers and statistical manipulations that has been used to provide scientific respectability for premises 1 and 2.

ANTECEDENTS OF THE "IQ ARGUMENT"

In 1855, Count Joseph Arthur de Gobineau published a four-volume *Essay on the Inequality of the Human Races*(11). De Gobineau concluded that there are six major races of people, and he based his case on the existence of six major classes of skin color. However, his case is confusing since he also ranks the races according to intelligence, strength, and beauty, which implies that intelligence, strength, and beauty can be used for classification purposes. According to de Gobineau, the white race possesses a monopoly on intelligence, strength, and beauty, and the black race is wanting in all three. "The negroid variety is the lowest and stands at the foot of the ladder. . . . His intellect will always move within a very narrow circle"(11; p. 205).

Even though de Gobineau used lots of words to defend his thesis, a defense was hardly needed at that time. His thesis was as compatible to western Europe and the United States in 1855 as was the view that the earth is flat 400 years earlier. He was not asked to define intelligence, nor to show how he had determined that whites have more of it than blacks, nor to defend the use of skin color and beauty as criteria for race classification. Uncritical acceptance of his thesis can be rationalized both from the purpose served by its support of foreign and domestic imperialism and by the soon to follow theory of evolution.

In 1859, Charles Darwin published the first of his major works explaining the origin of the species(12). His theory struck a blow at one of the predominant religious views of the time: special creation; this is sometimes counted as a victory of enlightenment over ignorance. But religion took it on the chin, regrouped, and continued its domination of people's minds pretty much as it always had. Yet, Darwin's theory of speciation did have a profound effect upon the dominant ideology, but in an unexpected way.

Not long after Darwin's first major publication, Herbert Spencer incorporated Darwin's theory of speciation into his own explanation of the rise and fall of civilizations, societies, and social institutions(13). Darwin's general statement said that existing species arise from preexisting species and that all modern species trace their origins to a common ancestor. He also proposed a mechanism by which species arise, a mechanism called *natural selection*. However, for natural selection to change the character of species, there must exist natural (genetic) variation within species. Spencer translated Darwin's concept of natural selection into the more descriptive phrases, "survival of the fittest" and "nature is red in tooth and claw"; to this translation, he added the authoritative and unscientific dictum that civilizations, societies, and social institutions compete for survival and that the *biologically fit* ones win.

Darwin based his theory of natural selection upon observation; he did not borrow it from physics or chemistry. The theory was designed to explain the particular phenomenon of speciation. Spencer used Darwin's theory to justify the racist conclusion that European societies are superior to Third-World societies and that capitalism is the most fit of all economic systems. Spencer's misuse of Darwin's theory went much farther; he said that the basis of fitness of civilizations, of societies, and of economic and political systems is *biological*. This notion extends the hereditarian concept of human nature quite beyond human nature and into the realm of higher-order phenomena embodied by social structure. There was not at the time, nor is there today, a shred of scientific evidence linking the genotypes of individuals to political, economic, and other social institutions.

Social Darwinism, as Spencer's thesis came to be called, is one form of biological determinism. The assumption of a hierarchy of civilizations in this thesis came from a variety of sources, de Gobineau and Western imperialism of the Third World included. From science came an even more potent force than de Gobineau: T. H. Huxley, one of the most influential of the nineteenth-century biologists. Huxley said in 1865: ". . . no rational man, cognizant of the facts, believes that the average Negro is the equal, still less the superior, of the average white man. And, if this be true, it is simply incredible that, when all his disabilities are removed, and our prognathous relative has a fair field and no favor, as well as no oppressor, he will be able to compete successfully with his bigger-brained and smaller-jawed rival, in a contest which is to be carried on by thoughts and not by bites"(14; p. 115).

At the time Spencer was building the case for social Darwinism, another Englishman, Francis Galton (Darwin's cousin), began his notorious work on "hereditary genius(15)," a work that included a genealogy of his own family, modesty notwithstanding. Among other things, Galton concluded that the upper classes of England possess the greater amount of inherited ability and, therefore, the biological privilege of being rulers and leaders (a clear example of the reification of an abstraction in support of political ideology). His observation that intellectual superiority runs in families led to the conclusion that intellectual capacity is inherited. This conclusion was not modified by the fact that wealth and opportunity also run in families.

Galton worked hard to find an objective measure of intelligence. Most of his efforts were expended upon attempts to find differences between dull and bright people through differences in reaction rates to physical stimuli. He acknowledged his failure in finding such a measure, but this did not deter him from concluding what his measure of intelligence was intended to prove—that black and working-class whites are less intelligent than upper-class whites. Galton proposed, based on this conclusion, that race crossing be forbidden since it would dilute the proportion of those with the highest intellect. Galton said in 1869 that Negroes are "two grades" lower than whites (as judged by intelligence) and that "the mistakes the Negroes made in their own matters were so childish, stupid, and simpleton-like, as frequently to make me ashamed of my own species"(15; p. 328).

Galton is called the "father of eugenics," the applied science of heredity concerned with the improvement of the human stock. He proposed that superior people breed prolifically (positive eugenics) and that inferior people be kept from breeding (negative eugenics) and that race crossing be forbidden. What better advice was wanted by a ruling class that had arrived at the same conclusion by way of "political analysis"?

Years later, eugenics got another shot in the arm from the work of David Starr Jordan(16), who studied pedigrees of the "unsuccessful." Jordan argued that criminality, feeblemindedness, poverty, perversity, and addiction are inherited traits, and his studies made famous the Jukes and Ishmalee families with their generations of poverty and antisocial behavior. Even though most of the works of Jordan have been discredited, the social attitudes left by them remain to this day.

In 1900, the papers of Gregor Mendel were discovered and his data were confirmed. The year 1900 is considered by many to be the birth date of the science of genetics.

Also in 1900, the minister of public instruction in Paris commissioned Alfred Binet to devise a test that would distinguish those school children who *could* but *were not* benefiting from education from those who *could not* benefit. The minister felt that some children who possessed learning potential were not living up to it and that, if such children could be identified, measures could be taken to help them. (This example illustrates the role of authorities in determining what scientists do and, at the same time, exposes the myth of the neutrality of science.)

The first intelligence test was made public in 1903(17). Binet searched for test items (questions and problems) that tested memory, the ability to discriminate, and vocabulary. He also constructed his tests so that the correct answers varied directly with the age of the children. That is, a test item was considered good if it was answered correctly more often by children of advancing age. A different set of items was developed for each age group between the ages of 6 and 16 years.

In 1912, Stern(18), of Germany, developed what is known today as the intelligence quotient, or IQ. IQ is determined by dividing a child's mental age by his or her chronological age. For example, each test item symbolizes 2 months of mental age. A 10-year-old child who solves correctly all items up to and including the 8-year-old items, four items in the 9-year-old set, three in the 10-year-old set, two in the 11-year-old set, and none in the higher set is said to have a mental age of 9.5 years (8 years + 18 months). This child has an IQ of 95 ($9.5/10 \times 100 = 95$). An 8-year-old child with the same mental age would be awarded an IQ of 120 by use of the same logic.

Binet did not provide a definition of intelligence, only a measure of scholastic potential. And, among its many uses, the intelligence test has been hailed more for its use as a predictor of scholastic success than for anything else. But this should not be surprising, since the tests were normalized against scholastic achievement and teacher expectation. Having found intelligence tests to be of some value as predictors of scholastic achievement, the enthusiasts now stretch the point to say that intelligence is what intelligence tests measure (premise 1).

If intelligence testing had become an academic exercise, it might have been possible to develop other measures and concepts of intelligence. But from the start the tests were designed to be used; that is, they were designed to separate children into groups based on learning capacity. The uses to which the tests were put led to an operational definition of intelligence, by-passing, as Jensen says, the need to define intelligence. And, it must be remembered, the tests were not confined to educational tracking. Consider their use in the eugenics movement.

In 1916, Terman, who later revised Binet's test into the Stanford-Binet test, said, "If we would preserve our state for a class of people worthy to possess it, we must prevent, as far as possible, the propagation of mental degenerates . . . the increasing spawn of degeneracy"(19).

The first eugenics law was passed in 1907, in Indiana. The preamble to that law began, "Whereas heredity plays a most important part in the transmission of crime, idiocy and imbecility. . . ." There are many important things to consider when examining this law, but let's look at only two. First, the science of genetics was only 7 years old at the time. Now, nearly 70 years later, the relationship between genes and poverty and idiocy and imbecility is still an enigma. Second, the first eugenics law was passed 4 years after Binet published the first intelligence test and 1 year before it was translated into English. Even though this test was designed to diagnose mental defectives, within 2 years it became the instrument with which to segment the population into idiot, imbecile, moron, feebleminded, subnormal, normal, above average, bright, very bright, and genius. Since all of these epithets had long since belonged to the culture, it can't be said that intelligence tests were the instruments with which they were *discovered.*

Four years later (1911), the New Jersey legislature added "feeble-mindedness, epilepsy, and *other defects*" to the list, and 2 years after that, Iowa added "lunatics, drunkards, drug fiends, moral and sexual perverts, diseased, and degenerate persons." By 1930, 31 states had enacted eugenics laws and more than 60,000 persons had been sterilized. In 1948, 27 states still had such laws. Not all states with the law used it, but some were extensive users. (No persons were sterilized under the eugenics law in Idaho, while about half of all sterilizations prior to 1930 were performed in California.)

In Germany, more than 56,000 persons were sterilized during the first year after passage of the Hereditary Health Law on July 13, 1933. Ultimately, 250,000 were sterilized under this law.

An investigating committee of the United States Senate, July 1973, headed by Edward M. Kennedy, obtained testimony from the Department of Health, Education, and Welfare that in 1972 at least 16,000 women and 8,000 men had been sterilized by the federal government. Of these, 365 were under the age of 21 and a high percentage of the 24,000 were black(20) (relative to the percentage of blacks in the population).

Between April 1972 and July 1973, there were 3,260 government-sponsored birth-control clinics. At least 100 minors were sterilized in these clinics during that time(20). In 1974, there were 14 states considering proposed legislation requiring women on welfare to submit to sterilization(20).

Based on IQ tests given to two million United States draftees during the First World War, it was estimated that roughly half of those tested were feebleminded. Following the war, the same tests were given to immigrants on Ellis Island. It was "discovered" that more than 80 percent of all eastern Europeans were feebleminded (89 percent of the Poles, 83 percent of the Jews, 87 percent of the Russians, etc.)(8). These data were used by eugenicists and geneticists to argue for passage of the Johnson Immigration Act (1924)(22). This act restricted immigration of eastern Europeans to the percentage of such people living in the United States in 1890, and the law was not repealed until 1965. (Note that these same IQ tests showed that eastern Europeans who had lived in the United States for at least 16 years were as bright as native-born citizens.)

Two of the primary motor forces of Hitler's fascist regime were the idea of superiority and the hereditarian concept of human nature (of course, this is fundamental to facism). As in his other endeavors, Hitler got support from science to do what he would have done without the support, that is, liquidate more than 6,000,000 Jews and communists. (When then Attorney General William Saxbe stated in 1974 that genes for communism tend to be more frequent in Jewish families, he was echoing the German geneticist, Lenz, who in 1931 provided Hitler with all the data needed to justify the hereditary improvement programs of the 1933-45 period. Hitler hated Jews, but he was even more apprehensive about communists. His geneticist advisors led him to the view that cleansing the earth of the former would eliminate the danger of the latter. Fascism is the classic method for using racism to terrorize and divide the working class[23].)

After the Second World War, there was an about-face with respect to eugenics. Nazi Germany illustrated for many that science could be subordinated to ideology and used to prove just about anything that one wants proved. However, even with the revulsion felt by many geneticists toward the Nazi atrocities, most would not have made their views public without the efforts of the United Nations to formulate a genetics manifesto: "As there is no reliable evidence that disadvantageous effects are produced thereby, no biological justification exists for prohibiting intermarriage between persons of different races. Available scientific knowledge provides no basis for believing that the groups of mankind differ in their innate capacity for intellectual capacity and emotional development"(24; pp. 14-15).

The UNESCO statement was ambiguous with respect to the effects of race crossing and somewhat positive in affirming that all races are essentially equal in intellectual potential. At least three prominent geneticists took issue with the first point, asserting that race crossing is potentially dangerous. Others opposed the second point, saying that, if differences among races are visible through physical and biochemical traits, there is every reason to expect that intelligence differences exist as well(22,25). Muller and Darlington continued efforts to expose the weaknesses of the UNESCO statement, but soon the race-intelligence-heredity furor died down. A few of the eugenics statutes were removed, while the others lay dormant. The years following the 1939-45 war were, so to say, the years of *social reforms.* (Eisenhower's heart attacks provided the stimulus for the massive expansion of Health, Education, and Welfare, and Kennedy's assassination greased the skids for Johnson's astonishing civil rights legislation; the policy makers began to act as if improvement of the environment is necessary to enhance the effectiveness and output of people.)

In January 1969, the environmentalistic approach to human betterment again came under attack by the hereditarians. I use the plural here because Jensen alone could not have created a ripple, let alone a splash. Jensen at the time was an obscure behavioral psychologist. His 1969 paper did not create a can of worms; it opened the can. Jensen's paper did not *create* support for its conclusions; rather, it opened the door for expression of preexisting support. Shortly afterwards, *The New York Times* gave the movement a name, "Jensenism"(26). But it is scientifically and politically incorrect to conclude that the movement is Jensen's or that he is a devil or a saint. Jensen is but another page in the long history of scientific racism; the path he followed was well marked before he had finished grammar school. His importance arises from the fact that his name symbolizes a bubble that would have burst with or without his prodding.

Let me illustrate this point. Hereditarians generally believe in school tracking systems, and a major instrument used to track students is the intelligence test. Jensen says that students who are genetically unable to learn ought not to be recipients of public monies that are used to provide superior educational services. Environmentalists generally believe that all students will benefit from superior education. Jensen's argument lends scientific weight to the conservative credo that rugged individualism will separate the sheep from the goats. In 1975, *Time* magazine summarized the controversy(27, p. 18):

There is probably no subject on which liberals and conservatives split more sharply than the causes and cures of crime. Liberals emphasize the unjust social conditions that are its breeding ground: slums, unemployment, poor education, racism, poverty amid affluence. . . . Conservatives are more apt to believe that deliberation, not desperation, is the root of crime. Says . . . former Attorney General William Saxbe, . . . "I believe a great many offenders commit crimes because they want to commit them." This disagreement is classic and deep. To conservatives, man has always been flawed by original sin—or simply by human weakness—but is in control of his own fate. To help him control it, the good society is obliged to emphasize a strong moral order, a respect for law and a confidence in punishment as a deterrent to crime.

The right-left split so permeates legal thinking that Walter B. Miller of the Harvard Law School's Center for Criminal Justice maintains that "ideology is the permanent hidden agenda of criminal justice." But ideological differences have recently started to blur under the impact of America's apparently permanent crime wave.

Since *Time* is neither radical nor left, the above statement cannot be served with anticommunism, as many liberal and left statements are, but this does not mean the statement is correct. I happen to think it is generally correct, as far as it goes, and that evidence of its general correctness is provided by the fact that the following major points of Jensen's 1969 paper were hailed so heartily by conservatives.

1. The environmentalists are wrong: compensatory education has failed because those for whom it was intended had previously reached the biological limits of their learning capacity.

2. The correlation between success in life and IQ arises mainly because intelligence (which IQ measures) is fixed by heredity and is, therefore, resistant to change through environmental manipulation. In fact, Jensen said, without evidence, that persons with low IQs are more resistant to change than persons with high IQs. Loosely translated, this means that there is no need to spend money trying to educate such people since success is not within their genetic grasp.

3. There is a wide variation of IQ scores within the white population of the United States, and about 80 percent of the variation is due to genetic differences among the individuals within the population.

4. The mean score of United States blacks is 15 IQ points below the white mean score, and the means are very different among social classes. (Jensen quotes data of Sir Cyril Burt that show mean IQ scores for six occupation classes, with high professional at the top [mean IQ equals 140] and unskilled laborers at the bottom [mean IQ equals 83]. These data have been severely criticized by Kamin[8]. Schwartz also discusses Jensen's data in another article in this volume.)

5. The most reasonable hypothesis to account for the black/white differences is that the differences are genetically determined.

These conclusions, together with the conservative concept of a genetically determined social structure, are *racist.* They are racist because they support the institutionalization of inequality based on race differences. It matters not at all whether Jensen is bigoted, prejudiced, or merely misled. What does matter is whether inequalities based on race are institutionalized. From this vantage point, it can be asked why so much time is spent quibbling with Jensen about whether the heritability of IQ test scores is 0.8 or 0.7 or 0.4. The heritability of IQ test scores is irrelevant, except to academics, if IQ scores do not measure intelligence or if IQ is not a major determinant of success. To be sidetracked by this kind of trivia when tens of millions of lives are repressed by institutional racism is to become an accomplice to scientific racism. However, it is instructive to review the distortions inflicted upon science by Jensen, Herrnstein, Shockley, Eysenck, and their followers, distortions which, coincidentally or not, support the hereditarian view of class structure.

A CLOSER LOOK AT THE IQ ARGUMENT

Let us examine in more detail the five major points of Jensen's argument, as presented above.

1. "IQ measures intelligence." Jensen is quoted as saying, "The notion is sometimes expressed that psychologists have misaimed with their intelligence tests. Although the tests may predict scholastic performance, it is said, they do not *really* measure intelligence—as if somehow the 'real thing' has eluded measurement and perhaps always will. But this is a misconception. We *can* measure intelligence. As the late Professor Edwin G. Boring pointed out, intelligence, by definition is what intelligence tests measure"(28; pp. 75-76).

Block and Dworkin(29) have shown the poverty of Jensen's position on the meaning of intelligence. At best, Jensen's is an operational definition of intelligence, and, at worst, it is a smoke screen. Here is a paraphrase of the Block/Dworkin argument: if an analogous definition of temperature were given (that is, if temperature is what thermometers measure) operationalism would commit its proponents to absurdities. First, such a definition tells us that thermometers cannot be improved upon and that no other device can measure temperature more accurately. Furthermore, if all thermometers malfunctioned to read 0°C, then everything would be the same temperature. However, if you say that nothing can measure intelligence better than an IQ test, it is possible that you are referring to an *ideal* test. If so, we can ask how closely real IQ tests approximate ideal ones, but this is the sort of question the operationalist is trying to avoid.

Another important point made by Block and Dworkin is that a definition of intelligence need not fix on the meaning of intelligence but could just as easily fix on its referent. A referent would succeed in providing a definition of intelligence that is the same for those who do and do not accept the definition. This still does not answer whether IQ tests succeed in measuring intelligence, or anything else, and for this reason a referent isn't attractive to operationalists. The operational definitions of Jensen fix meaning, not referent.

One of the maddening aspects of blatant operationalists, and this is directed at Jensen and Herrnstein, is their insistence that measurements of intelligence can proceed without a theory of intelligence. Block and Dworkin ask what should a theory do, and they answer(29; pp. 339-40):

It should explain the causal role of intelligence in phenomena in which intelligence has a causal role. For example, it should explain how intelligence affects learning, problem solving, understanding, discovering, explaining, etc. Also it should explain how factors which affect intelligence do so. A good theory should say something about what intelligence is and what people who differ in intelligence differ in (information processing capacity? memory?) though this latter task would presumably be a by-product of the former tasks.

The following quote from Jensen points out his atheoretical stance(28; p. 72):

Disagreements and arguments can perhaps be forestalled if we take an operational stance. First of all, this means that probably the most important fact about intelligence is that we can measure it. Intelligence, like electricity, is easier to measure than to define.

Herrnstein, in "I.Q.," follows suit with(3; pp. 48-49):

Binet invented the modern intelligence test without saying what intelligence is. . . . Some rough and ready notion of intelligence lurked in the background—having to do with mental alertness, comprehension, speed, and so on—but he was not forced to defend an abstract definition in order to sell the ideas of his test to the world. Instead, he could point to how well the test worked. Rarely did a bright child, as judged by the adults around him, score poorly, and rarely did a poor scorer seem otherwise bright . . . in general most children ended up about where they were expected to.

In other words, Herrnstein is saying that the problems attendant on the construction of theory can be side-stepped if IQ scores correlate with intelligence. Block and Dworkin offer the following analogy(29; p. 341):

You notice that when it rains you develop an ache in your knee. Its strength is proportional to the amount of rainfall. . . . You then realize you can measure rainfall merely by attending to the severity of the pain in your knee. There is a fatal flaw involved in supposing this picture applies to measuring intelligence; you cannot measure intelligence by finding items which correlate with it unless you already have a way of measuring it.

In the case of rainfall, there is another way of measuring it. In the case of Herrnstein's defense of Binet, the "other measure" is "the adults around him." But even the avid psychometrician Terman, who developed the Stanford-Binet test, voiced opposition to personal judgment as a means for standardizing IQ tests. (If all the judges agree [providing a reliable measure], they still may not be accurate.)

Regardless of the importance one attaches to theory and the development of theory by measurement, it is correct to say that there is no definition of intelligence based on theory. In this way, the science of intelligence testing differs from physics, chemistry, and biology. Again, according to Block and Dworkin(29; p. 342):

Jensen's comparison of intelligence with electricity is instructive. What would a scientist be measuring if he measured electricity independently of theories about it? Electric potential (measured in volts)? Electric current (amps)? Electrical power (watts)? Electrical work (joules)? Electrical charge (coulombs)? . . . The fact is that "electricity," like any serious theoretical term is easy to define, if one knows the theory. For example, "electricity" can be defined as the motion of electric charges, and its various aspects can be measured on the basis of the theory which licenses the definition.

One last word on the meaning of intelligence. After Jensen goes through a lengthy discussion defending an atheoretical approach to the subject, reminding us that intelligence is what intelligence tests measure, we find that his descriptive uses of the term revert back to popular language and the emotional associations that have existed within the culture for hundreds of years. The population variation, from low to high IQ, is described with the terms, "smart," "stupid," "bright," "dull," "intelligent," and "unintelligent." In this, Jensen parallels earlier workers by being more a slave to his racist culture than to scientific objectivity, as exemplified by the following quotation from Terman, Yerkes, and Cattell:

Their dullness seems to be racial, or at least inherent in the family stocks from which they come. The fact that one meets this type with such extraordinary frequency among Indians, Mexicans, and Negroes suggests quite forcibly that the whole question of racial differences in mental traits will have to be taken up anew and by experimental methods. The writer predicts that when this is done, there will be discovered enormously significant racial differences in intelligence, which cannot be wiped out by any scheme of mental culture(19; pp. 91-92).

The report takes the above results to indicate that the negro as compared with the white man of equal intelligence is relatively strong in the use of language, in acquaintance with verbal meaning, and in perception and observation; and that he is relatively weak in judgment, in ability to analyze and define exactly, and in reasoning(21; p. 738).

The main lines are laid down by heredity—a man is born a man and not an ape. A savage brought up in a cultivated society will not only retain his dark skin, but is likely to have also the incoherent mind of his race(30; p. 165).

2. "IQ is a causal determinant of success." It is correct that IQ test scores and success in school and life are positively correlated. However, there is no evidence that they are causally related, as Jensen and Herrnstein claim. No doubt the following

points from Herrnstein's article in *The Atlantic Monthly* will illustrate both the quality of his science and the conclusions drawn from it(3; pp. 58 and 63):

1. *If differences in mental abilities are inherited, and*
2. *If success requires those abilities, and*
3. *If earnings and prestige depend on success,*
4. *Then social standing (which reflects earning and prestige) will be based to some extent on inherited differences among people.*

The fact that IQ scores and success are correlated does not mean that IQ causes success. IQ/success correlations are in fact *built into* the test by the methods of standardization and by "score causation." Methods of standardization include a variety of success standards, such as the testimonies of teachers. Score causation—the fact that high-scoring students are viewed differently than low-scoring students—includes teacher expectations and the assignment of students to "tracks," creating a lesser or greater likelihood of school achievement.

Bowles and Gintis(7) have shown that IQ is a notoriously poor predictor of success in any given job and that many factors unrelated to IQ are more highly correlated with success (for example, the income of the parents, the number of years of schooling, and the geographical location of the family home). S. E. Luria put the case squarely on the line(31; p. 27):

The son of an industrialist with an IQ of 90 has an enormously better chance to succeed than a black child with 120. . . . Whenever self-appointed experts state that the problem of impoverishment of IQ is a major problem facing our nation, I see racist eugenics raising again its ugly head. Behind the urgent scientific necessity to know the truth about those miserable 15 IQ points, (between black and white mean scores) on which the whole future of the schools, the nation, and the species is claimed to depend: there is a movement to drop the current efforts toward integrated schools and equalized opportunities for black and white children.

Block and Dworkin(29) state that the history of IQ testing right up to the present is littered with the corpses of tests that were dropped because they failed to correlate sufficiently with measures of success. Extant tests are those which have passed the correlation hurdles.

Probably the most insidious of the arguments that IQ is causally related to success is that of Richard Herrnstein. He tries to build a case for the belief that socioeconomic standing (SES) is a biological imperative. His argument goes like this: if you take a random sample of people from each of the social classes and ask them to rank all of the current professions in order of prestige, all will produce the same ranking. Truck driving and garbage collecting are always near the bottom, and physics, medicine, and law are always near the top. Therefore, he says, the SES scale is the externalization of inherent wisdom. He doesn't even bother to argue against the most obvious alternative hypothesis that SES standings have been internalized (through propaganda, rewards and punishment, and direct observation). As Clarence Kailin notes(32; p. 3): "Most of our history has been written by—and to suit the needs of—a propertied, segregated, white male-oriented society. Until recently our texts have ignored the very existence of black people (and all other racial minorities), except as delightfully happy and contented slaves—passive, docile, imitative and childlike—concepts based on the racist myth that Africans and their descendants are biologically inferior."

Almost any study of history will show that our prejudices are historically conditioned. One of these preconditioned misconceptions is that white supremacy is natural, normal, and instinctive. And, because this implies genetic origins, it means that our attitudes and prejudices are predetermined and, therefore, must always be with us. Is it any wonder, then, that measures of success will be standardized by success? Ever since Binet published the first intelligence test, good test items have been those a good student got right and a poor student got wrong.

3. "IQ is highly heritable." Heritability is the proportion of a characteristic's variation in a population that is due to genetic differences among the individuals within the population. The concept of heritability as used by Jensen is based on the general concept that the total phenotypic variation (V_P) of a measurable characteristic (e.g., IQ scores) can be partitioned into variance due to environmental factors (V_E) and variance due to genotypic differences among individuals within the population (V_G). This relationship tells us, then, that $V_P = V_E + V_G$. In the case of IQ scores, this means that variation within the real population is due to both genetic and environmental inputs. The task is to isolate either V_E or V_G; the other can then be calculated since $V_E + V_G = 1$.

The problem of estimating heritability is not as simple as this relationship leads one to believe, (as Schwartz's article shows), and a *good* estimate of heritability would not tell us anything about the relationship between genes and characteristics. On this latter point consider the following: let us determine the heritability of the number of kidneys per person in the population of United States college students. As you know, the vast majority of college students have two kidneys, which means that the variance within the population is small. A few people will have one kidney, and still fewer will have three; those with one will have had the other removed surgically, and those with three will have had a transplant. Now, if all persons with one or three kidneys had had surgery, then all the variation within the population would be due to environmental forces, V_E. Therefore, V_G equals 0. Does the fact that V_G equals 0 mean that genes are not involved with kidney development (including number, size, shape, location, etc.)? No. A heritability that is low or zero means only that the phenotypic variance in the population due to genetic differences among individuals is low (it means only that the people are genetically alike with respect to the characteristic in question). Clearly, the heritability of a trait does not tell us whether or not the trait is "genetically determined." Heritability tells us about the variance within populations, not the relationship between genes and phenotypes.

A more esoteric obfuscation is to misinterpret the covariance and interaction of environmental and genotypic inputs to phenotype. If the genetic advantage of a child, for example, appears as early maturation (sometimes called precosity), there exist two obvious outcomes as determined by intelligence tests. One, early maturation will lead educated, middle-class white parents to shower rewards and incentives for continued intellectual progress. Better and better environments for learning will be provided, and soon the child or young adult will begin creating his/her own environments on the basis of the reward system that designed the concept of intelligence in the first place. Would it be correct to attribute that part of the total variance due to covariance between genes and environment to V_G or to V_E? Jensen favors adding it to V_G, not because it belongs there, but because that gives him a higher heritability estimate.

On the other hand, precosity sometimes bugs people, and, rather than bring reward, it initiates punishment. At a very early age, precosity can track a child toward isolation or toward "antisocial" means of getting attention. It is instructive to examine the rather modern concept of hyperactivity. In the good old days, when class sizes were small, active children often were considered to be bright and they got more than their share of the teacher's attention, which in turn improved their chances of becoming bright (Pygmalion effect). Today, with large class sizes, active children can't depend upon encouragement from overworked teachers; to the contrary, they are a nuisance. Thus, their behavior is often associated with low IQ. In other words, two identical genotypes can lead to very different phenotypes by covarying with different environments.

In addition to ignoring or displacing covariance and interaction between genotype and environment, the data used to construct high heritability estimates are biased. One of the most popular ways to estimate heritability within human populations is to use twin studies. Identical twins, for example, should be more alike (concordant) than unrelated persons if the measured trait is due to genetics. The extent to which they differ is said to be a measure of the environmental influence upon phenotype. But the means for collecting the data are often biased. For example, Kamin(8) uses the data of Newman, Freeman, and Holzinger, who studied 19 pairs of identical twins, to illustrate how sampling procedures bias results. These 19 pairs of twins were found through a nationwide advertising campaign during which questions were asked in order to determine with certainty that the respondents were identical twins. The costs of housing the twins in Chicago were high, so the investigators did not want to risk having to care for a large number of nonidentical twins. Twins that were in some ways different were screened from the study before they were ever examined for identity, leaving, of course, built-in concordance between twin pairs. Kamin concludes from a number of biases found in the data that there is a good case for concluding zero heritability for IQ scores. However, as Block and Dworkin remind us, rather than to estimate zero heritability from worthless data, it is correct to conclude that no estimate can be made.

There are two sides of the heritability coin that are relevant to Jensen's main theme. The first is the claim that, within the white population, the heritability of IQ-score variation is high (which there is no reason to believe, given the difficulty of estimating heritability in human populations and the worthlessness of existing data). Jensen assumes from this that high heritability also characterizes the variation in the black population. But the second point goes beyond this assumption to argue that, given high heritability with the two populations, the most reasonable explanation for the difference between them is that they are genetically different. The logic of this argument is fallacious(10).

The absurdity of the conclusion that "there are no 'black genes' or 'white genes;' there are intelligence genes, which are found in populations in different proportions, somewhat like the distribution of blood types. The number of intelligence genes seems lower, overall, in the black population than in the white"(26; p. 43) puts a completely new twist on premise 4, which states that black and lower-class individuals' mean IQ scores are lower than middle- and upper-class white individuals' mean scores. There is *no* evidence that the mean differences are due to genetic differences, and there is every reason to suspect that the mean differences reflect nothing more than differential access to upward mobility.

All human beings have thousands of genes, probably at least as many experiences, and a phenotype that changes through time. Characteristics of phenotype that are influenced by genes are also influenced by experience. Indeed, genes do different things in different environments; but what is often lost in discussion of genes and experience is that the temporal sequence of experiences is as important to the outcome as the quality of the experiences. The interacting mechanisms and their effects on phenotype are unknown. Yet, in the tradition of authority, which enlightenment was reputed to have slain, Jensen, Shockley, Herrnstein, and their less well known supporters will argue that human phenotypes can be fractionated in such a way as to show which parts are shaped by genes and which by experience. Such an approach to human nature will always call for the "good society" to compensate for nature's mistakes, and, of course, the ultimate compensation for the kinds of mistakes Jensen is talking about is governmental control over the lives of all people, first by educational tracking, then by employment, and finally by repression of minority opinion.

It is not the point of my argument to find flaws in Jensen's arithmetic or to suggest better ways of estimating heritability. The important issue is the role of the IQ argument within the context of science, academia, and our racist society. To complete this analysis, it is necessary to say a few words about science in relationship to the dominant ideology (see reference 1 for a fuller statement).

SCIENTIFIC RACISM

The struggle against racism and oppression of minority and working-class people cannot be won by better heritability estimates; the struggle cannot be won by science. The issue is political, and it will be won or lost in political struggle.

Evidence for this position comes as much from the right as from the left. For example, the first congressional committee on un-American activities under the direction of Representative Hamilton Fish, in 1930, was given the authority to investigate "communism in the United States" (at the time, it was advertised that communists had provoked the Great Depression). The committee's report said, in part, "The Committee recommends the enactment of federal law to prosecute Communists in spreading false rumors for purpose of causing runs on the banks. . . . Communism is a world-wide political organization advocating *absolute racial equality*"(33; p. 37). [My emphasis.] On January 14, 1976, the House transferred the same authority to the Judiciary Committee. The next day, the Committee reviewed a legislative proposal that is probably the most repressive ever considered by a House committee, a 753-page revision of the United States Criminal Code, known as Senate Bill No. 1 or S-1:

*Not since the Alien and Sedition Act has a more sweeping assault been mounted in this country against democratic self-government (*Los Angeles Times, *May 18, 1975, part IX, p. 2).*

*Fear feeds upon itself and becomes the nation's own worst enemy, creating an atmosphere in which some are willing to sell their birthright for a police state. A bill now before the U.S. Senate is a sign that that kind of disease did not disappear with the departure of Richard Nixon from the White House (*Atlanta Journal, *February 25, 1975, p. 15A).*

Probably the most difficult lesson for middle-class Americans to learn is that United States capitalists depend for their survival upon racism. This runs all the way from maintaining cheap and divided labor forces, to high rents for run-down housing, to saving money for defense by cutting medical, educational, and other social services. While Johnson waged war on poverty and talked of "The Great Society," Daniel Moynihan wrote a major government report to prove that the appalling oppression against which black people had fought was caused not by capitalist society, but by inherent deficiencies in the black family structure(34). Following this came Jensen's assertion that black people are genetically inferior to whites. Then, Edward Banfield(35) asserted that oppressed workers like to live in poverty, enjoy slums, and have a sociological predilection for beating their children.

The same week that Richard Nixon announced phase one of the wage freeze, Herrnstein announced, in *The Atlantic Monthly,* that the capitalist society is a genetically determined meritocracy and that wealth and status vary directly with intelligence. He went so far as to say that unemployment is inherited in the same manner that bad teeth are inherited. Following this, Bane and Jencks(36), with the help of a $500,000 grant from the Carnegie Foundation, published a study that implied that current and future massive budget cuts in the schools are justified—because the basic determinants of success are "luck and personality." All of these racists are publicized in the media, hailed as courageous spokesmen for unpopular "scientific discoveries," and zealously defended when their "free speech" is trod upon by outraged antiracists.

There is a common theme in all these "scientific" assertions: they all blame the victim as a screen to cover imperialist acts of genocide, profit-making wars, and crimes that are possible only when fortified by incredible power. Could Jensenism have become a household word without the backing of the ruling class and their managers who control the media? Could Shockley have gotten untold numbers of free hours on national and local television to promote a blatantly genocidal proposal to sterilize the victims of ruling-class greed without the backing of the ruling class?

Examine for a moment Shockley's "logic" and why it is so appealing to the ruling class. He proposed legislation calling for the sterilization of persons with low IQs with a cash inducement of $1,000 per IQ point below 100. Legislation is not debated by poor people. Just the same, Shockley says(4):

Eugenics is a shunned word because it was a feature of Hitlerism. But the lesson of Nazi history is not that eugenics is intolerable. . . . [He uses Denmark as a good example of eugenics.] The real lesson of Nazi history was anticipated 140 years before Hitler, when the Bill of Rights incorporated into our Constitution the First Amendment guaranteeing freedom of speech and of the press. Only the most anti-Teutonic racist can believe the German people to be such an evil breed that they would have tolerated the concentration camps and gas chambers if a working First Amendment had permitted exposure and discussion of Hitler's final solution—the extermination of the Jews.

Is Shockley saying here that we need not fear his proposals for sterilization because the First Amendment will allow us to debate openly whether those of us judged to be "inferior" are to be castrated or gased? (It is not unrelated to this point that the S-1 bill boils down to a repeal of the First Amendment.)

The two basic themes upon which scientific racism is founded are that blacks and other non-whites are born with inferior brains and a limited capacity for mental growth and that the personalities of non-whites are abnormal, either by nature or by nurture.

The Committee against Racism was established for the expressed purpose of building a movement to oppose scientific racism. Shortly after its founding, it became obvious to everyone in CAR that scientific racism is but one tentacle of academic racism and that academic racism is but one arm of a multiarmed attack on *all* working-class people by the ruling class. Scientific racism dresses up cultural prejudice and bigotry in esoteric terminology and statistical jargon, and academic racism peddles the product for a ruling class, which then translates the results into institutional racism. The only solution to this kind of oppression is to build a multiracial, rank-and-file movement that grows through reform work. Since the reforms that are needed cannot and will not be acceded to, it will become clear to those who still hold to "the great American dream" that the ruling class has *no* intention of serving any but its *own* interests.

Therefore, racism is part and parcel of the class struggle. Racism is one of the major tools used by the ruling class to divide and conquer the working class. And racism has served the interests of the ruling class sufficiently well that they will fight to the death to maintain it. For this reason, CAR's campaign against racism is buttressed by the reality that racism hurts everyone (except the ruling class and its loyal managers) and that the political struggle to eliminate racism must become a class struggle.

I have argued that science reflects and shapes characteristics of the society of which it is a part. A dominant characteristic of society is institutional racism, and I have illustrated that its feedback loops with science. Another characteristic of society is the existence of a ruling class, the class that owns and controls the societal institutions, science included(37). As ideology gives rise to a popularized philosophy that masks societal-scientific relationships, so too there is a popularized philosophy that masks the relationship between the ruling class and the lower classes of society. The first step to any reform struggle is to understand the masks behind which the real forces in society hide.

References

1. Woodward, V. 1974. The ideology of science: A view from Vietnam. *The American Biology Teacher* 36:21-27, 87-92.
2. Jensen, A. R. 1969. How much can we boost IQ and scholastic achievement? *Harvard Educational Review* 39:1-123.
3. Herrnstein, R. J. 1971. I.Q. *The Atlantic Monthly* 228(3):43-64. [See also, 1973. *I.Q. in the meritocracy*. Boston: Atlantic-Little, Brown and Company.]
4. Shockley, W. 1972. Dysgenics, geneticity, and raceology. *Phi Delta Kappan* LIII(5):305.
5. Neary, J. 1970. A scientist's variations on a disturbing racial theme. *Life* 68(22):58B-65.
6. Daniels, N. 1975. IQ, heritability and human nature. *Proceedings of the Philosophy of Science Association.*
7. Bowles, S., and Gintis, H. 1972-73. IQ and the U.S. class structure. *Social Policy 3* (4–November-December 1972–and 5–January-February 1973).
8. Kamin, L. 1974. *The science and politics of I.Q.* New York: Halsted Press.
9. Feldman, M. S., and Lewontin, R. C. 1976. The heritability hang-up. *Science* 190:1163-68.
10. Lewontin, R. C. 1970. Race and intelligence. *Bulletin of the Atomic Scientists* 26:2-8.
11. de Gobineau, J. A. 1967. *The inequality of races.* Trans. Adrian Collins. New York: Howard Fertig.
12. Darwin, C. 1859. *On the origin of species by means of natural selection, or the preservation of the favored races in the struggle for life.* London: John Murray.
13. Spencer, H. 1897. *The principles of sociology.* New York: D. Appleton and Company. [See also, Hofstadter, R. 1959. *Social Darwinism in American thought.* New York: Houghton Mifflin Company.]
14. Huxley, T. H. 1865. Emancipation–Black and white. In *Lectures and lay sermons,* ed. Ernest Rhys, pp. 115-20. London: J. M. Dent and Sons.
15. Galton, F. G. 1869. *Hereditary genius: An inquiry into its laws and consequences.* New York: Macmillan.
16. Jordan, D. S. 1922. *The days of a man: Being memoirs of a naturalist, teacher, and minor prophet of democracy.* Vol. 1. New York: World Book Company.
17. Binet, A., and Henri, W. 1903. *L'ètude experimentale de l'intelligence.* Paris: Schleicher Frères.
18. Stern, W. 1912. *Die Psychologische Methoden der Intelligenzprufung.* Leipzig: Barth.
19. Terman, L. M. 1916. *The measurement of intelligence.* Boston: Houghton Mifflin Company. [See also, 1917. Feeble-minded children in the public schools of California. *School and Society* 5(111):165.]
20. Aptheker, H. 1974. Sterilization, experimentation and imperialism. *Political Affairs.* January 1974, p. 37.
21. Yerkes, R. M. 1921. Psychological examining in the U.S. Army. *Memoirs of the National Academy of Sciences,* vol. 15, p. 655.
22. Allen, G. E. 1975. Genetics, eugenics and class struggle. *Genetics* 79(supplement):29-45.
23. Weinreich, M. 1946. *Hitler's professors.* Yivo, N.Y.: Yiddish Scientific Institute.
24. UNESCO. 1952. *The race concept.* Paris: UNESCO.
25. Provine, W. B. 1973. Geneticists and the biology of race crossing. *Science* 182:790.
26. Edgon, L. 1969. Jensenism. *The New York Times Magazine,* August 31, 1969, pp. 10-11, 40-47.

27. *Time*. 1975. The crime wave. *Time* 105:10-14, 18.
28. Jensen, A. R. 1973. *Genetics and education*. New York: Harper and Row.
29. Block, N. J., and Dworkin. G. 1974. IQ: Heritability and inequality, parts 1 and 2. *Philosophy and Public Affairs* 3:331-409 and 4:40-99.
30. Cattell, J. M. 1947. *James McKeen Cattell: American man of science*. Vol. II. Lancaster, Pa.: Science Press.
31. Luria, S. E. 1974. What can biologists solve? *New York Review of Books* 21(1):27-28.
32. Kailin, C. 1974. Teach the history of racism. *The CAR Wheel* 1(2):3-13. [Available through the Committee against Racism, 5742 Dogwood Place, Madison, Wis. 53705.]
33. Wilkinson, F. 1975. From HUAC to S-1. *The Center Magazine* VIII:35-43.
34. Moynihan, D. P. 1967. *The Negro family: A case for national action*. Cambridge, Mass.: MIT Press.
35. Banfield, E. C. 1970. *The unheavenly city*. Boston: Little, Brown and Company.
36. Bane, M. J., and Jencks, C. 1973. Five myths about your IQ. *Harper's* 246:28.
37. Domhoff, G. W. 1967. *Who rules America?* Englewood Cliffs, N.J.: Prentice-Hall.

Suggested Readings

Bodmer, W. F., and Cavalli-Sforza, L. L. 1970. Intelligence and race. *Scientific American* 223:19-29.

Chase, Allan. 1977. *The legacy of Malthus: The social costs of the new scientific racism*. New York: Alfred A. Knopf.

Daniels, N. 1973. Smart white man's burden. *Harper's* 247:24-40.

Garcia, J. 1972. IQ-The conspiracy. *Psychology Today* 6(4): 40-43, 92-94.

Gossett, T. F. 1963. *Race: The history of an idea in America*. Dallas: Southern Methodist University Press.

Koestler, A., and Smythies, J. R. 1969. *Beyond reductionism*. Boston: Beacon Press.

Layzer, D. 1974. Heritability analyses of IQ scores: Science or numerology? *Science* 183:1259.

Lewontin, R. C. 1974. Analysis of variance and analysis of causes. *American Journal of Human Genetics* 26:400.

Ludmerer, K. 1972. *Genetics and American society*. Baltimore: Johns Hopkins University Press.

Progressive Labor Party, 1974. *Racism, intelligence and the working class*. [Available by writing to Box 808, Brooklyn, N.Y. 11201.]

Science for the People. 1974. IQ. *Science for the People*, Spring 1974.

Sex Roles

It is a common assumption in our society that men and women have different roles because they are biologically different. With the growth of biology in the nineteenth century, these differences in roles were explained by scientists who related them to specific anatomical and physiological differences. Menstruation, pregnancy, nursing, smaller body size, and hormonal differences were among the factors that set women apart and fitted them naturally into their role as mother and housewife. These factors also incapacitated them for various other kinds of work.

More recently, psychological differences between men and women have been thought to have hormonal or other biological bases. Lionel Tiger, in his book *Men in Groups*(1), claims that bonding between men is basic to the organization of human society and that the inability of women to form lasting bonds with other women explains their low participation in business, politics, and other leadership roles. Tiger ascribes the importance of male-male bonding to our evolutionary history as hunter-gatherers. Males, being stronger, went out to hunt large animals in groups, while women had to remain close to home and care for the children. Thus, selection produced males able to cooperate effectively to form an efficient hunting group. Hunting also selected for such traits as aggressiveness and competitiveness. The hunter-gatherer model of the evolution of human sex roles has been elaborated upon in the symposium *Man the Hunter*(2) and critiqued by Sally Slocum in an article entitled "Woman the Gatherer: Male Bias in Anthropology"(3).

Views that explain modern sex roles as natural products of evolution have recently come under attack with the rise of the women's movement and its struggle for equality. Are women's roles constrained by their biology, or are we taught our gender by society through the expectations of our parents, teachers, and peers? The far-reaching political implications of this question make it likely to remain controversial for many years.

References

1. Tiger, L. 1969. *Men in groups.* New York: Random House.
2. Lee, R. B., and DeVore, I., eds. 1969. *Man the hunter.* Chicago: Aldine.
3. Slocum, S. 1975. Woman the gatherer: Male bias in anthropology. In *Toward an anthropology of women,* ed. Rayna Reiter. New York: Monthly Review Press.

Science and Sex Roles in the Victorian Era

Robin Miller Jacoby

As an historian specializing in women's history, I have become increasingly aware that certain aspects of the history of science provide an important perspective for understanding social attitudes regarding sex roles. In an attempt to illustrate some of the links between scientific and social views, this paper will examine what I see as the mutually reinforcing relationship that existed between nineteenth-century biological theories and nineteenth-century ideology on sex roles. My hope is that this paper, in conjunction with the others in this book, will increase awareness of the fact that scientific research is not carried out in an ideological vacuum. Those of us contributing to this volume are firmly committed to the importance and the value of various types of scientific research, but we also think it extremely important that all of us—scientists and nonscientists—be aware that the social context in which scientists live and work often influences the topics they choose to research, and, concomitantly, the results of their research often have significant impact on their society.

By way of introduction, I will begin with a few brief, general comments on the notion of sex roles and an explanation of why I have chosen the Victorian era as the period of focus for this paper. Following that, I will summarize Victorian ideology regarding sex roles, and I will then be able to examine the links between these cultural norms and nineteenth-century scientific theories.

Sex roles have characterized every human society known to scholars. However, aside from the biological constant that women have been the only sex to bear children, the work of historians and anthropologists has revealed that, at different times and in different cultures, very different sets of roles and attributes have been ascribed to the sexes. In other words, the *concept* of sex roles has been a constant throughout human history, but the *content* of those roles has varied considerably. An excellent illustration of the varying content of sex roles is found in Margaret Mead's *Sex and Temperament in Three Primitive Societies*(1). In the early 1930s, Mead did fieldwork among three primitive tribes living within a 100-mile area in New Guinea. She found that in one tribe both the men and the women conformed to our stereotypes of female behavior; that is, both sexes were mild mannered and nurturing. In the second tribe, the dominant pattern was for men and women to be what we consider masculine, aggressive, tough, and militaristic. Whereas in the third tribe, she found a society based on what we would consider a reversal of sex roles, "hard" women and "soft" men. Mead's book is important because it indicates that many of the differences between men and women that we assume to be basic and "natural" are, in fact, the products of cultural conditioning rather than biology.

Since the Victorian period in British and American history was a time of considerable preoccupation with definitions of sex roles among social thinkers and scientists, it is a particularly rich and relevant one for examining connections between cultural norms and scientific theories. Sex roles were the subject of such widespread concern and discussion because one of the social implications of the economic changes involved in industrialization and urbanization was the development of new patterns of living and working for all people, but especially for middle-class women. I'll return to this point, but for now I want to mention a few other characteristics of the Victorian period that link it to our own.

As Lewontin points out in the opening paper in this volume, a widespread interest in science and the scientific method was a dominant intellectual current of the nineteenth century. He mentions some of the significant advances in scientific knowledge that occurred during this period and talks about the explicit consciousness of Victorian intellectuals that they were living in a world rapidly changing due to major upheavals in the social, economic, political, and scientific spheres. Victorian intellectuals were anxious to understand the new world in which they were living, and it was this desire to produce rational explanations of the physical and social structure of their society that led to such increased interest in the natural sciences and to the development of what we now refer to as the social sciences, academic disciplines that attempt to understand the social sphere through use of scientific methodology. Thus, it was during the Victorian era that science and the scientific method were imbued with the kind of authority and respect that we now take for granted. For many Victorians, scientific explanations came to replace theological ones as the primary authorities on the physical and social world, although, as some of the quotations in this paper indicate, scientists were not always averse to using religious supports for some of their propositions on sex roles.

A further reason for focusing on the Victorian period is that the views on sex roles that held the status of "official ideology" during that time are still very much with us today. Thus, the Victorian period is linked to our own in that it was the first historical era during which significant numbers of people lived in urban, industrialized settings; it shared our basic belief in science and the scientific method as a rational approach to understanding the physical and social world; and many aspects of Victorian views on appropriate spheres and behavior for women and men exist as powerful forces within our own period.

Victorian ideology regarding sex roles was strongly influenced by various social and economic changes that resulted from industrialization, and, as I mentioned earlier, these changes had particular significance for the lives of middle-class women. In the preindustrial period, women of all classes tended to play a major role in economic production. They generally performed tasks different from those of men, but the tasks of both women and men were crucial for the economic sustenance of the family. Men were typically the producers and women the processors of raw materials needed for food and clothing. For example, a common pattern was for males to be responsible for raising sheep and the females for turning the sheared wool into cloth and then into clothing. Another characteristic of preindustrial life was that work tended to take place within the home and on the land surrounding it. With industrialization, the processing of raw materials was increasingly removed from the home to the factory, and among the first processes to be industrialized were those related to the production of textiles and clothing, processes that had traditionally fallen within the sphere of work done by women. Women had either done this work directly for their own families or, in wealthier families, had been responsible for supervising female domestic servants in these tasks. With the removal of such work from the home, working-class women simply followed their work from the home to factories or workshops. Middle-class women, however, found their economic role shifting from being involved in the production of goods to becoming consumers of factory-made products. The growing availability of commercially produced products, coupled with the increased supply of domestic servants (due to immigration and urbanization), meant that bourgeois women could now become "ladies of leisure."

The cultural norm throughout the Victorian period for families living in urban areas was that women, especially married women, worked only if absolutely necessary. It was in this period that the notion developed that it reflected negatively on a man's role as a provider if his wife or daughters worked outside the home. Despite the number of working-class women who were members of the labor force, this cultural norm that women, particularly married women, should not work was pervasive throughout American society. Obviously, not every family could afford to adhere to this ideal, but, on the basis of my research on nineteenth-century working-class women, I am convinced that it was an ideal that was widely accepted on a cross-class basis; women became members of the labor force only when it was essential for their family's survival.

To compensate middle-class women for their decreasing involvement in economic production and domestic work, increased emphasis began to be placed on other traditional aspects of the role of women within the home. An enormous quantity of fictional and nonfictional literature that discusses appropriate roles and behavior for women was produced during the early years of the Victorian period. This body of prescriptive literature codified a social ideology known today as "the cult of true womanhood"(2). To be a "true woman" was considered the "natural" inclination of women. In practice, it meant being pious, pure (chaste in mind and demeanor and not sullied by involvement in the world of politics and business), submissive (to father, brothers, and husbands), and domestic (living a life focused on home and family). It was assumed that leading the life of a "true woman" would be totally fulfilling for women because it represented a healthy social expression of their biological dictates. It was posited that women who did not exhibit these qualities of piety, purity, submission, and domesticity—or who explicitly rejected them—were unnatural creatures whose behavior was a threat to the social order and contrary to the will of God. As the author of a book published in 1841 and entitled *Woman in America: Being an Examination into the Moral and Intellectual Condition of American Female Society* wrote(3; p. 144 [Cott]):

Our chief aim throughout these pages is to prove that her domestic duties have a paramount claim over everything else upon her attention—that home *is her appropriate sphere of action; and that whenever she neglects these duties, or goes out of this sphere of action to mingle in any of the great public movements of the day, she is deserting the station which God and nature have assigned to her.*

This concept of the station that nature and God had assigned to women raises the question of scientific views regarding sex roles, and, as the rest of this paper will demonstrate, the scientific community emphatically reinforced the cultural ideology that men belonged actively and aggressively in the world and that women belonged passively and submissively in the home. Scientists upheld this rigid division of sex roles, claiming that simple observation and rigorous, sophisticated research proved that women were physically and intellectually inferior to men.

This perception of woman as the weaker sex was based on three "facts." The first was the accurate observation that females tended to be physically smaller than males, which led to the not entirely justified conclusion that women were, therefore, physically inferior to men. Although there may be a positive correlation between physical size, musculature, and physical strength, there is a significant difference between a society that encourages women to develop their bodies to reach something approaching their physical potential and a society that emphasizes the notion that women are weaklings and that it is unfeminine to be otherwise. Scientists also posited that women had less stamina than men, and female proclivities to fainting were cited in support of this view. It is clear to us now (and was clear to some people then) that fashionable clothing styles were the primary cause of the high incidence of female fainting during the Victorian period. The tightly laced corsets that women wore were the main culprit in producing female weakness, for years of corset wearing tended to deform women's diaphragms to such an extent that many of them literally could not breathe properly. (See the illustrations in Sklar[4], pp. 174, 209-10, and 213.) The third aspect of women's lives that was cited in support of this notion of female frailty was their menstrual cycle, for women were considered to be seriously incapacitated several days out of every month due to menstruation. (See the article by Bart in this volume.)

As a result of these perceptions of women's physical nature, scientist after scientist propounded the view that the only life-style females could tolerate was a sheltered one within the home. Involvement in the world outside the home was simply more than women could handle physically. It is extremely important to be aware of the class bias in this point of view. Working-class women were not considered too delicate to do hard physical labor in factories or at home, and domestic servants were never excused from their duties because they were menstruating. In other words, these notions of female frailty were strongly tied to the concept of what is meant to be a lady, and the option of adhering to this aspect of ladylike behavior was denied to working-class women.

The assumption that women were intellectually inferior to men was based upon numerous scientific studies that established that females had smaller brains than males and upon the unjustified conclusion that there was a correlation between brain size and intellectual ability.[1] (These same studies and the same erroneous interpretations of data were used to "prove" that blacks were intellectually inferior to whites.) A typical statement reflecting the views of leading scientists in the area of brain research was, "The grown up Negro partakes, as regards his intellectual faculties, of the nature of the child, the female, and the senile White"(5; p. 192 [p. 51]).

Nineteenth-century scientists spent endless hours measuring the cranial capacities of males and females, and they concluded that there was an average difference of about 220 cubic centimeters between the sexes(6; p. 49). More recent studies suggest an average difference of about 150 cubic centimeters. However, regardless of the degree of difference, there are two problems with moving from those data to the conclusion that females are intellectually inferior to males. The first is that, in humans, brain size is correlated with body weight and height, and comparisons of male and female cranial capacity that fail to take these critical factors into account are meaningless. (Because women are, on the average, lighter and shorter than men, it is reasonable to expect

[1] My analysis of this Victorian research on the correlation between brain size and intellectual ability is based on discussions with Professor Leonard Radinsky of the University of Chicago department of anatomy.

their brains to be smaller in absolute terms.) The second is that there has never been any evidence to indicate that there is a correlation between normal human brain size–either absolute or relative–and intellectual ability. The conclusion of ninteenth-century scientists, who blissfully ignored the need to question their assumptions about the relationship between cranial capacity and intellectual ability, were summarized in the statement of one eminent researcher that "woman is a constantly growing child, and in the brain, as in so many other parts of her body, she conforms to her childish type"(6; p. 54). These scientists, who lived in cultures where the social roles of women were considered inferior to those of men, moved unhesitatingly and with great assurance from scientifically accurate collections of data on cranial capacity to interpretations of those research data that were, in fact, more reflective of prevailing social ideology than of pure, unbiased scientific reasoning.

Scientific views on sexuality formed another important component of authoritative literature on sex roles. It is important to keep in mind that this literature was prescriptive and not necessarily descriptive of Victorian sexual behavior. In other words, this literature can tell us how leading Victorian scientists thought people should behave, but it cannot be taken as evidence of actual behavior patterns. However, given the authority accorded to scientific views in this period and the widespread publicity these ideas received, I think it is reasonable to assume that this prescriptive literature did have a significant impact on Victorian behavior(7).

Dominant scientific views on sexuality were predicated on three widely accepted tenets. The first was that the human body contained a limited amount of energy (or vital force, to use the Victorian term) and this finite resource had to be utilized carefully so that one area of activity did not drain away energy needed in other areas. The second assumption was that sexual instinct was the basest, most primitive human emotion. This moral view was reinforced by scientific demonstrations of what scientists called *amativeness* (the source of sexual instinct), which was said to be located at the base of the brain and was not among what they considered to be the higher, more complex convolutions of the cerebrum. The third component of Victorian theories on sexuality was the assumption that sexual feelings in males were very strong but were virtually absent in females, at least in ladies. Conventional wisdom of the period was that males were faced with a constant and intense struggle between their passions and their reason; while they were not encouraged to give in to their passions, it was not expected that they could always resist them. (This is one reason why the use of prostitutes was so widely condoned in the Victorian era[8].)

The combination of these three tenets was responsible for producing the following set of views on sexuality among leading Victorian scientists. Puberty, the time of physiological maturation, was considered a crisis period for both sexes. It was defined as the time when boys should become strong and vigorous and girls frail, timid, and modest. Puberty was considered a particularly traumatic period for females, for, if they did not develop the proper physiological and emotional equilibria at that stage, it would affect not only the rest of their lives but also those of their children. (This attitude reflected Victorian views on heredity, which assumed that most personality traits were inherited rather than acquired characteristics.) Since females were considered to possess a lesser amount of vital force than males and since it was recognized that the female reproductive system was more complex than the male's, it was held that females should be especially careful to channel their limited energy into the proper areas; the most important area needing energy was, of course, the female's reproductive system. Assuming that a woman was truly fulfilled only through motherhood, doctors insisted that virtually all of an adolescent female's energy was needed to insure her reproductive potential.

As a corollary of this point of view, scientists led the resistance against feminist demands that women be allowed access to higher education, seriously arguing that it was an inappropriate pursuit for young women because their brains and their ovaries could not develop simultaneously. Scientists, particularly physicians, claimed that women who attended college achieved whatever intellectual development they did at the expense of the healthy functioning of their reproductive systems. An influential and widely read statement of this point of view was *Sex in Education* by Dr. Edward Clarke, a faculty member of Harvard Medical School(9). This book was published in the mid-1870s and was an immediate best-seller. It consisted of a series of case histories from Clarke's practice, all of them concerning young women who had been perfectly healthy prior to attending college but who then found that their breasts had ceased to develop, that they had begun to have extremely painful menstrual periods, or that they had stopped menstruating altogether.

Complementing this biological argument was the assumption that females were intellectually inferior to males. Opponents of higher education for women argued that young women were intellectually incapable of handling the same rigorous education offered to young men. Academicians joined in this aspect of the debate, claiming that it would be a waste of their time and talents to try to educate young women; there was also great concern that allowing women access to degree programs would cause a lowering of academic standards.

According to standard Victorian medical opinion, the focus of a female's life from puberty to marriage should be on the healthy development of her reproductive organs, and doctors advocated a routine of bed making, cleaning, and child tending as appropriate activities to prepare young girls for their future roles as wives and mothers. In addition, doctors urged females to avoid strong emotions, especially anger, to get plenty of fresh air and moderate exercise, and to wear proper clothing, which was the one issue on which Victorian doctors and feminists were in agreement.

As I indicated earlier, menstruation was regarded as a periodic illness to which females were subject. One very popular book, which was originally published in 1832 and sold thousands of copies throughout the nineteenth century, set the tone for scientific attitudes toward menstruation. Its author wrote, "During its [menstruation's] continuance, the woman is said to be unwell or out of order. . . . Indigestible food, dancing in warm rooms, sudden exposure to cold or wet, and mental agitations should be avoided as much as possible"(10; p. 24 [p. 39]). Victorian attitudes toward menstruation are an excellent example of the influence of theories of biological determinism on theories regarding sex roles, for, in this case, a basic biological fact of female existence was laden with enormous and pernicious social implications. A typical expression of Victorian attitudes toward menstruation is the following statement made by James McGregor Allen in a paper given at the 1869 meeting of the Anthropological Society of London(11; p. 40 [Vicinus]):

Although the duration of the menstrual period differs greatly according to race, temperament, and health, it will be within the mark to state that women are unwell from this cause, on the average two days in the month, or say one month in the year. At such times, women are unfit for any great mental or physical labour. They suffer under a languor and depression which disqualify them for thought or action, and render it extremely doubtful how far they can be considered responsible beings while the crisis lasts. Much of the inconsequent conduct of women, their petulance, caprice, and irritability, may be traced directly to this cause. It is not improbable that instances of feminine cruelty (which startle us as so inconsistent with the normal gentleness of the sex) are attributable to mental excitement caused by this periodic ill-

ness. . . . Michelet defines woman as an invalid. Such she emphatically is, as compared with man. . . . In intellectual labour, man has surpassed, does now, and always will surpass woman, for the obvious reason that nature does not periodically interrupt his thought and application.

Even if a woman somehow survived the educational process with her reproductive system intact, scientists contended that it was out of the question that she could function as a competent professional, especially as a scientist. As one doctor wrote, "One shudders to think of the conclusions arrived at by female bacteriologists or histologists at the period when their entire system, both physical and mental, is, so to speak, 'unstrung,' to say nothing of the terrible mistakes which a lady surgeon might make under similar conditions"(12; p. 319 [p. 59]).

While the socialization of adolescent girls was to prepare them for their roles as future mothers, the socialization advocated for adolescent boys was to train them for economic success. The Victorian era was a period of enormous economic development, and a prevalent notion was that there were great opportunities for upward mobility. It was very much the era of the self-made man, and it was held that, to succeed economically, men must concentrate their energy on getting ahead in the material world. However, males had to be very careful, for their strong sex drive posed a serious threat to their need to focus their energies on training that would provide the basis for future economic success. Semen was considered the most potent vital force of the male body, for it was believed that 1 ounce of semen was equivalent to 40 ounces of blood(6; p. 97). Given this theory, sexual activity was clearly an enormous drain on the rest of the body's ability to function vigorously. In addition, it was claimed that excessive sexual activity led to premature decay of the male body.

The connection between economics and sexuality was reflected in the Victorian vocabulary. Sexual continence was talked of as an equivalent of thrift, and the term for an ejaculation was an *expenditure.* In colloquial language, ejaculating was talked of as spending. Analogies were often made between the need and opportunity for men to conquer themselves (to exert rational control over their irrational sexual passions) and the economic and technological conquest of nature. Terms associated with mining and railroad building, two very important areas of nineteenth-century economic activity, were filled with sexual overtones. Going into further detail on this topic is beyond the scope of this paper; the general point is that economic development was discussed in language that made it seem a socially valued and virile outlet for sublimated sexual impulses(13).

The concern over adolescent males was not that (like females) they could not do things such as attending college because it would inhibit the development of their reproductive system; it was focused instead on the impulses toward sexual activity that resulted from a male's physical maturation. Victorian writings on sexuality were filled with discussions of the harmful consequences of masturbation, which was viewed as an evil, debilitating practice. Given the Victorian's opinions about limited energy and the potency of semen as a vital force, what can only be described as their obsession with masturbation becomes somewhat understandable, although irrational nonetheless. A typical statement of Victorian attitudes is found in a book by William Acton, a prominent Victorian urologist, who described an habitual adolescent masturbator in the following terms(14; p. 19 [Marcus]):

The frame is stunted and weak, the muscles undeveloped, the eye is sunken and heavy, the complexion is sallow, pasty, or covered with spots of acne, the hands are damp and cold, and the skin moist. The boy shuns the society of others, creeps about alone, joins with repugnance in the amusements of his school-fellows. He cannot look anyone in the face, and becomes careless in dress and unclean in person. His intellect has become sluggish and enfeebled, and if his evil habits are persisted in, he may end in becoming a drivelling idiot.

Stephen Marcus, the author of a very interesting book on sexuality in the Victorian period, points out that such descriptions, which had the status of "official knowledge," raise very real questions about the relationship between belief and behavior. Marcus posits(15; pp. 19-20):

We can reasonably assume that masturbation was practiced among adolescents to about the same extent then as it is now—that is to say, it was as good as universal (Acton himself admits this). And we know too that most adolescents did not correspond to Acton's description of the masturbator, and that the largest part of them grew up to be what for want of a better word we must call normal males.

Taking Victorian adolescent males and females into the next stage of their life cycles, we find that attitudes on sexual activity within marriage provide a further example of connections between scientific thought and social norms regarding sex roles. As I indicated earlier, a prevalent view maintained toward male and female sexuality was that the sexual drive was natural for men but more or less absent in women. In the words of an article in a highly respected Victorian journal, "In men in general the sexual drive is inherent and spontaneous and belongs to the condition of puberty. In the other sex, the desire is dormant, if not non-existent"(16; pp. 456-57 [p. 82]). According to Dr. Acton (in his 1871 publication [14; pp. 112-13]), "The majority of women (happily for them) are not very much troubled with sexual feeling of any kind"(8; p. 82). After allowing for aberrations such as nymphomania, which he considered a form of insanity, Acton indicated that he had no doubt(8; p. 83):

that sexual feeling in the female is in the majority of cases in abeyance . . . and even if roused (which in many instances it never can be) is very moderate compared with that of the male. . . . Many of the best mothers, wives, and managers of households, know little of or are careless about sexual indulgences. Love of home, children, and of domestic duties are the only passions they feel. As a general rule, a modest woman seldom desires any sexual gratification for herself. She submits to her husband's embraces, but principally to gratify him; and were it not for the desire of maternity would far rather be relieved from his attentions.

Just in case a woman happened to discover that sex was an activity she enjoyed, another doctor warned women that "voluptuous spasms" during coitus were certain to interfere with conception, which was, of course, the only reason why women should be inclined to engage in sexual intercourse(17; p. 256 [p. 101]).

Even though men were supposed to have such high-powered sex drives, it was recognized by many writers that there was considerable public ignorance on sexual functioning. One book on sexual hygiene had the following case history(18; p. 27 [Walters]):

So dense indeed is the general ignorance on this important subject that it is not an uncommon event for men and women to enter the married state without any preliminary knowledge whatever, either theoretical or practical, of the sexual relation. It is not long since that an educated man, a patient, told me that when married at the age of twenty-five he was totally unaware of the nature of the sexual relation so far as his part of it was concerned. His wife unfortunately shared the same ignorance. After a few days or rather nights spent in the midst of curious sensations and doubts and fears they obtained the information in some indirect way that the male organ should be introduced into the "water passage" of the female. The gentlemen essayed the experiment on several different occasions without success, and was then compelled to the ludicrous necessity of searching for the place with a candle. Even then it was some days before he succeeded in effecting a natural intercourse, and that too after other consultations with friends who were more versed in the matter than they were. This ignorance, of course, is more generally confined to women than to men, but there are enough instances among the latter, to make it a matter of surprise even to a physician who ought to be accustomed to surprises.

Although less guilt was attached to sexual activity within marriage, restraint was still considered important because excessive sex was seen as a drain on men and an imposition on women. A typical example of advice on the topic of sexual frequency was(19; p. 87 [Walters]):

It is . . . impossible to lay down a precise rule, which will be equally adapted to all men, in regard to the frequency of their connubial commerce. But as a general rule, it may be said to the healthy and robust, it were better for you not to exceed, in the frequency of your indulgences, the number of months in the year; and you cannot habitually exceed the number of weeks in the year, without in some degree impairing your constitutional powers, shortening your lives, and increasing your liability to disease and suffering—if indeed you do not thereby actually induce disease of the worst and most painful kind, and at the same time transmit to your offspring an impaired constitution with strong and unhappy predispositions.

Thus, medical theory held that intercourse once a month was quite sufficient, although it is not at all clear how many couples actually practiced abstinence to this extent. (One problem with this view is that scientists did not understand correctly the relationship between ovulation and fertility and assumed that women were most fertile just before and just after menstruating.)

A final quote on the topic of sex within marriage, which is rather interesting from a feminist point of view, is this statement that dramatically points out the price women paid for having to submit to excessive sexual activity(20; p. 85 [Walters]):

The husband, in the exercise of what he is pleased to term his "marital rights," places his wife, in a very short time, on the nervous, delicate, sickly list. In the blindness and ignorance of his animal nature he requires prompt obedience to his desires, and, ignorant of the law of right in this direction, thinking that it is her duty to accede to his wishes, though perhaps fulfilling them with a sore and troubled heart, allows him passively, never lovingly, to exercise daily and weekly, month in and month out, the low and beastly of his nature, and eventually slowly but surely, to kill her. And this man, who has as surely committed murder as has the convicted assassin, lures to his net and takes unto him another wife, to repeat the same program of legalized prostitution on his part, and sickness and premature death on her part.

These scientific views had significant social consequences. They were responsible for sustaining an ideal of sexual repression for both men and women that we now consider emotionally damaging. In addition, they provided "objective" and authoritative reinforcement for a rigid social ideology regarding appropriate roles for men and women. Men were faced with pressures to achieve economic success in the world outside the home, while the restriction of women to the domestic sphere was seen as consistent with the state of their intellects and the dictates of their bodies.

The full weight of scientific authority was brought to bear against Victorian feminists, who were struggling, not for the sexual liberation of women, but for expanded education, economic, political, and legal rights and opportunities for women. Scientists disputed feminist claims to equality, arguing that the reforms they advocated were unnatural, upsetting to the social order, and contrary to the will of God. Concluding an address on sexual equality to the Pennsylvania Medical Society in 1882, the president of that organization stated, "It may be safely asserted that any doctrine or scheme of reform . . . which conflicts with the plain teachings of the revealed will of God is ruinous to the present happiness and ultimate destiny of mankind"(22; pp. 26-27 [p. 75]). Another prominent doctor wrote, in an article published in an 1893 issue of the *New York Medical Journal,* "Any woman who promoted equality of the sexes had either given evidence of masculo-feminity or had shown conclusively that she was a victim of psycho-sexual aberrancy"(6; p. 77).

Many of these Victorian attitudes toward sex roles and sexuality sound absurd today. It is difficult for us to believe that serious, intelligent, highly respected scientists could propound such theories and that people would believe them. Yet, more careful reflection leads to the realization that many of these attitudes are still prevalent today, albeit in subtler, more muted versions.

As in the Victorian era, scientific research today carries great authority and commands considerable respect, which means that scientists are in the position to have significant influence on social theories. The explicit point of this paper was to point out the role of the Victorian scientific community in reinforcing social values. The paper's implicit message was to make those of you who are part of the twentieth-century scientific community aware of how much work remains to be done to sort out what is cultural from what is "natural" in the roles and potentials of men and women. Even more important, I hope that the next time any of you reads about or participates in a scientific study whose conclusions support traditional ideas on sex roles you will stop and think very carefully about what cultural assumptions may be underlying the research and affecting the researcher's interpretations of the data.[2]

References

1. Mead, M. 1963. *Sex and temperament in three primitive societies.* New York: Dell.
2. See Welter, B. 1966. The cult of true womanhood: 1820-1860. *American Quarterly* XVIII (Summer):151-74.
3. Graves, A. J. 1841. *Woman in America: Being an examination into the moral and intellectual condition of American female society.* New York: Harper and Brothers. [Excerpted in N. F. Cott, ed. 1972. *Root of bitterness: Documents of the social history of American women.* New York: E. P. Dutton and Company.]
4. Sklar, K. K. 1973. *Catherine Beecher: A study in American domesticity.* New Haven, Conn.: Yale University Press.
5. Vogt, C. 1864. *Lectures on man, his place in creation and in the history of the earth.* London: Anthropological Society. [Quoted in J. S. Haller and R. M. Haller. 1974. *The physician and sexuality in Victorian America.* Urbana, Ill.: University of Illinois Press.]
6. Haller, J. S., and Haller, R. M. 1974. *The physician and sexuality in Victorian America.* Urbana, Ill.: University of Illinois Press.
7. For a discussion of the relationship between this prescriptive literature and female sexual behavior, see Degler, C. N. 1974. What ought to be and what was: Women's sexuality in the nineteenth century. *American Historical Review* LXXIX(5):1467-90.
8. See, for example, Sigsworth, E. M., and Wyke, T. J. 1972. A study of Victorian prostitution and venereal disease. In *Suffer and be still: Women in the Victorian age,* ed. Martha Vicinus. Bloomington, Ind.: Indiana University Press.
9. Clarke, E. H. 1875. *Sex in education.* Boston: J. R. Osgood and Company.
10. Knowlton, C. (no date) *Fruits of philosophy.* New York: Truth Seeker Company. [Quoted in E. Showalter and E. Showalter. 1972. Victorian women and menstruation. In *Suffer and be*

[2]In addition to the secondary works and documentary anthologies cited in the body of this article, the references numbered 22 through 27 in the reference list were useful in the preparation of this article. I would also like to thank Margaret Lourie and Leonard Radinsky for their help in the preparation of this paper.

still: Women in the Victorian age, ed. Martha Vicinus. Bloomington, Ind.: Indiana University Press.]

11. *Anthropological Review.* 1869. *Anthropological Review* VII: cxviii-cxix. [Quoted in E. Showalter and E. Showalter. 1972. Victorian women and menstruation. In *Suffer and be still: Women in the Victorian age,* ed. Martha Vicinus. Bloomington, Ind.: Indiana University Press.]
12. Irwell, L. 1896. The competition of the sexes and its results. *American Medico-Bulletin* X:319. [Quoted in J. S. Haller and R. M. Haller. 1974. *The physician and sexuality in Victorian America.* Urbana, Ill.: University of Illinois Press.]
13. See Barker-Benfield, G. J. 1973. The spermatic economy: A nineteenth century view of sexuality. In *The American family in social-historical perspective,* ed. Michael Gordon. New York: St. Martin's Press.
14. Acton, W. 1871. *The functions and diseases of the reproductive organs, in childhood, youth, adult age, and advanced life, considered in their physiological, social and moral relations.* Philadelphia: Lindsey and Blaikston. [Quoted in S. Marcus. 1974. *The other Victorians: A study of sexuality and pornography in mid-nineteenth century England.* New York: Basic Books.]
15. Marcus, S. 1974. *The other Victorians: A study of sexuality and pornography in mid-nineteenth century England.* New York: Basic Books.
16. *The Westminster Review.* 1850. Pages 456-57. [Quoted in E. M. Sigsworth and T. J. Wyke. 1972. A study of Victorian prostitution and venereal disease. In *Suffer and be still: Women in the Victorian age,* ed. Martha Vicinus. Bloomington, Ind.: Indiana University Press.]
17. Walker, A. 1839. *Intermarriage: Or the mode in which and the causes why beauty, health, and intellect result from certain unions, and deformity, disease, and insanity from others.* New York: J. and H. G. Langley. [Quoted in J. S. Haller and R. M. Haller. 1974. *The physician and sexuality in Victorian America.* Urbana, Ill.: University of Illinois Press.]
18. Howe, J. W. 1883. *Excessive venery, masturbation and continence: The etiology, pathology and treatment of the diseases resulting from venereal excesses, masturbation and continence.* New York: Bermingham and Company. [Excerpted in R. G. Walters, ed. 1974. *Primers for prudery: Sexual advice to Victorian America.* Englewood Cliffs, N.J.: Prentice-Hall.]
19. Graham, S. (no date) *Chastity, in a course of lectures to young men.* New York: Fowler and Wells. [Excerpted in R. G. Walters, ed. 1974. *Primers for prudery: Sexual advice to Victorian America.* Englewood Cliffs, N.J.: Prentice-Hall.]
20. Cowan, J. 1880. *The science of a new life.* New York: Cowan and Company. [Excerpted in R. G. Walters, ed. 1974. *Primers for prudery: Sexual advice to Victorian America.* Englewood Cliffs, N.J.: Prentice-Hall.]
21. Ziegler, J. L. 1882. Address—Woman's sphere. Medical Society of the State of Pennsylvania, *Transactions* XIV. [Quoted in J. S. Haller and R. M. Haller. 1974. *The physician and sexuality in Victorian America.* Urbana, Ill.: University of Illinois Press.]
22. Cominos, P. T. 1972. Innocent femina sensualis in unconscious conflict. [In *Suffer and be still: Women in the Victorian age,* ed. Martha Vicinus. Bloomington, Ind.: Indiana University Press.]
23. Rosenberg, C. 1973. Sexuality, class and role in nineteenth century America. *American Quarterly* XXV(May):131-53.
24. Smith, D. S. 1973. Family limitation, sexual control, and domestic feminism in Victorian America. *Feminist Studies* I(Winter-Spring):40-57.
25. Smith-Rosenberg, C. 1972. The hysterical woman: Sex roles and role conflict in nineteenth century America. *Social Research* XXXIX(Winter):652-78.
26. Smith-Rosenberg, C. 1973. Puberty to menopause: The cycle of femininity in nineteenth century America. *Feminist Studies* I(Winter-Spring):58-72.
27. Wood, A. D. 1973. The fashionable diseases: Women's diseases and their treatment in nineteenth century America. *Journal of Interdisciplinary History* IV(Summer):25-52.

Biological Determinism and Sexism: Is It All in the Ovaries?

Pauline B. Bart

Why is the question of biological determinism and sex roles an issue, as it is in the two quotations that follow? It is an issue because, if one assumes that our gender behavior is determined by our biology and furthermore that biology is immutable, then it follows that traditional gender behavior cannot be changed.

In other words, the testis of the developing male fetus superimposes masculinity on a basically neutral or female fetus. A castrated male fetus will develop in the same way as the female fetus, regardless of whether or not her ovaries are present, emphasizing not only the dominant role of maleness, but also the completely passive role of the fetal ovary(1; p. 56).

In conclusion, how can we attempt to summarize all that has been said in this chapter? In all systems that we have considered, maleness means mastery; the y-chromosome over the x, the medulla over the cortex, androgen over oestrogens. So physiologically speaking, there is no justification for believing in the equality of the sexes; vive la différence!(1; p. 70).

The use of biology to justify sexism and racism is a recurrent theme in our intellectual tradition. Since we know from the sociology of knowledge that the intellectual products of an era tend to reflect the perspectives and interests of those in power, we can predict that biological explanations of behavior would characterize conservative eras, while environmental explanations would characterize liberal eras. Thus, it is no accident that community psychiatry is on the decline, while biogenetic explanations and pharmacological treatment of mental illness are now where it's at in these more conservative seventies. Tavris' research(2) shows that belief in biological determinism is correlated with conservative views regarding social change. Even if change occurs, it is at best transitory, according to people believing in biological determinism. Lionel Tiger and Joseph Shepher's recent study of women in the kibbutz(3), for example, takes this position.

(I should note that biological and sociocultural explanations are not necessarily either/or. They do interact. Moreover, it looks as though biology is not always the independent variable. On occasion, the biology follows from the social arrangements;

for example, preliminary findings indicate that, after a monkey achieves a dominant position in the hierarchy, his testosterone level rises [4].)

An additional noteworthy fact about the invocation of biological explanations to account for sex roles is that they are invoked differentially; that is, they are used to explain female behavior but not male behavior. Only women are "menopausal," are bitchy because "it's that time of the month," have penis envy, and so on. Apparently, biology affects women more than men, but only negatively. As Parlee(5) states, such explanations, attributions, and, as sociologists say, vocabularies of motives take attention away from the women's *situation* that could account for her symptoms and moods. In the medical and psychiatric literature that is the source of much of the writing about female reproductive processes, either the psychiatric aspects are considered epiphenomena of the physical functions or, as the Lennanes have noted(6), genuine physical discomfort that women experience is considered "just psychological" and not treated because the pain simply stems from the women's failure to accept her natural feminine role (e.g., in menstrual cramps, pregnancy, and labor).

Moreover, the meaning of biological factors is provided by the society. For example, one's height, hair and eyes, and body type are primarily determined genetically. Yet, the meaning of these factors depends on what particular form of beauty is valued by a society. The life chances and the interactive patterns of a woman in our society who is a tall, slender, blue-eyed, curvaceous blonde are different from those of a short, heavy, dark woman.

However, although Freud was wrong in saying that anatomy is destiny, anatomy is not irrelevant. For example, self-examination of one's cervix with the use of a plastic speculum is an important part of the women's health movement, as a demystification technique. It testifies to the importance of the difference between female and male anatomy, since, were female genitalia external as are male, it would not be necessary. The concern in the women's movement for rape testifies to the fact that, while it is almost impossible for women to rape men, it is all too easy for men to rape women or other men.

In this paper, first, I will show how science in the nineteenth century operated in the interest of the *status quo,* legitimizing sexism as well as racism with its sacred mantle; second, I will show by example how biological determinism remains important in modern medicine; and, third, I will conclude with the question of the kibbutz—how and why women's roles went, as Rae Blumberg puts it, from liberation to laundry(7).

HISTORICAL PERSPECTIVE

It is helpful when examining sex roles and their determinants to look at the historical background. The image of the middle-class nineteenth-century woman held by her physician, and thereby disseminated to the society, was that of a uterus and ovaries surrounded by various peripheral accoutrements. Every thought, every feeling, every action, every inaction, and every illness was attributed to the vicissitudes of those powerful organs. She, and society, could ignore them only at her, and its, peril. Higher education, for example, would handicap those organs for childbirth. Full attention had to be paid to their care and development. It was a zero-sum equation—what the brain got, the uterus, ovaries, and white Western civilization lost. It followed, therefore, that, when women exhibited "unfeminine" behavior such as masturbating or wanting a career or when they suffered from any other "disease," the solution, by the end of the century, had become the removal of her "diseased" ovaries, from which these disturbances stemmed. Some historians consider this behavior on the part of gynecologists, who had just driven out the midwives, a response to the first wave of the women's movement and the anxieties it aroused in the male population(8).

When physicians tore their attention away from the reproductive tract, it fell on the female brain. "Only in admitting her true nature in the home circle and then seeking respectability within that circle could woman avoid the misfortunes of neurasthenia"(9; p. 33). Her brain was ill equipped to withstand urban life, bridge playing, eating lunch out, education, and strong sexual desire. "Beware!! Science pronounces that the woman who studies is lost"(10; p. 238 [p.33]).

It is interesting and important to note that the same doctors who advised a domestic life for women used arguments for women's inferiority almost identical with those who rationalized racism. For example, according to Haller and Haller, Harry Campbell, M.D., noted that "just as black children showed an early precocity and atrophy, so the white woman's mental evolution appeared to blossom early, frequently achieving bright success during her early years, then declining after a short span of time" in contrast to the white male who developed more slowly but "continued to mature beyond puberty"(9; p. 38). The woman, like the Negro, "had innate limitations on her capacities." For example, both lacked originality. In addition, medical "scientists" noted similarities between women and Negroes, both being flat-footed and more susceptible to infantile illnesses of the "lower races," such as anemia. The anthropometricists measured brains and found that, among Negroes and Australian aborigines, there were few differences by sex, but, as the ladder of civilization was climbed (culminating in white Anglo-Saxon gentlemen), the difference increased, a demonstration of white male superiority.

Brain research is the most patent example of science in the interest of ideology. When researchers demonstrated that the female brain was lighter than the male's but was heavier in proportion to body weight, craniometricists were forced to take another look at their instruments, becoming "a medium through which medical science, in its desire for certainty, evaluated and attacked the demands of the 'new woman'. In pursuit of an alternative craniometric tool, Manouvrier turned to an ingenious yet questionable index which related brain weight to weight of the thigh bone"(9; p. 51). Comparing brain weight of men and women of the same height, Boyd, Broca, and Bischoff assigned the male a brain-weight (intellectuality) advantage of 12 percent, 9 percent, and 11 percent respectively. Here, they felt, lay sound confirmation of male superiority. A fitting quote with which to close this section is the following from another physician in 1895: ". . . neither the 'emancipated woman' at one end of the scale nor the prostitute at the other propagates her kind, and society has reason to be thankful in both cases"(11; p. 978 [p. 56]).

BIOLOGICAL DETERMINISM IN MODERN MEDICINE

Another dramatic example of the way in which biological arguments have been instrumental in promoting and reinforcing sexism has been gynecology textbooks. A context analysis of such textbooks over time demonstrates this(12):

Feeble and sensitive at birth, and destined by nature to give us existence and to preserve us afterwards by means of her tender and watchful care, woman, the most faithful companion of man, may be regarded as the very complement of the benefits bestowed upon us by the Divine Being; as an object fitted to excite our highest interest and presenting to the philosopher, as well as to the physician, a vast field of contemplation.

Whereas before puberty she existed but for herself alone, when all her charms are in full bloom, she now belongs to the entire species which she is destined to perpetuate by bearing almost all the burden of reproduction.

During the forties, the Second World War gave women an opportunity to climb up from the Victorian pedestal. Rosie the Riveter could not be considered "feeble and sensitive." A textbook author responded as follows(13; pp. 59-60):

The very recent widening of the sphere of feminine activities, with the assumption of the male function of protection and maintenance has led to a further weakening of the reproductive urge, resulting in the modern "smart" type—sexless, frigid, self-sufficient.

He continues:

The fundamental biologic factor in women is the reproductive urge of motherhood balanced by the fact that sexual pleasure is entirely secondary or even absent. One of the commonest problems presented for solution by the gynecologist is the vast and fundamental difference between the sexes in regard to sexual appetite. Women with their almost universal relative frigidity *are apt to react to the marital relationship in one of three ways: (a) They submit philosophically to their husbands; (b) They submit rebelliously as a matter of duty; (c) They rebel completely and through refusal try to force the husband to adapt himself to their own scale of sexual appetite. [My emphasis.]*

Male sexuality, of course, is more important and of a much higher nature. Thus, Cooke continues, "Biologically for the preservation of the race, the male is created to fertilize as many females as possible and hence is given an infinite appetite and capacity for intercourse." (He never heard of the refractory period?) More recent textbooks show surprisingly little change. In 1962, Parsons stated: "Some women are truly frigid . . . psychic factors operate at the level of the cerebral cortex to inhibit the translation of sexual stimuli into a pleasurable response. Unless there is a true aversion to sex, the marital relations may proceed without *disturbing* either partner"(14; p. 494). [My emphasis.] (You can sleep through the whole thing, in other words.) And, in 1971, Dr. Thomas Green of Harvard tells us(15; p. 436):

If the sexual inadequacy on the part of the wife stems from a fundamental immaturity and inability or failure to assume the normal adult female role in the marital relationship, he (the gynecologist) may be able to help by gradually imparting to her the nature of what her role should be—as Sturgis has described it so well, the fact that although [sic] the instinctive sexual *drive of the male, who carries the primary responsibility for biologic survival of the race, is greater than hers. [My emphasis.]*

Thus, she has to make herself available for the fulfillment of his needs.

After reading some of these statements from gynecology textbooks that demonstrate how "science" is used to support the *status quo,* it should not surprise us to learn that the fact that women have menstrual cycles is also used by "scientists" to serve an ideological function—keeping women in traditional roles. What is said about female cycles has political importance. It was not uncommon to read that menstruation was the uterus weeping for the child that was not safely implanted in its walls. It was only a few years ago that Edgar Berman, Hubert Humphrey's physician, announced that women could not hold important political positions because they were subject to "raging hormonal influences"(16; p. 35). Women's menarche and menstruation are never ignored. Usually, menstrual blood is considered taboo, and, in some cultures, menstruating women are ostracized lest they contaminate the rest of the tribe. In our society, the potential for contamination is handled somewhat more subtly. It is her *psychic* state that can cause the problems to others, and it is her fear of making her shameful state public that contaminates *her.* I will discuss briefly some alternative approaches to the female reproductive cycle, approaches that consider or emphasize sociocultural factors rather than focusing on biology.

Paige(17) explains all the affective fluctuations during the menstrual cycle in terms of culture. She finds that women of different ethnic groups (e.g., Jewish, WASP, Catholic) differ in their affective fluctuation, and she accounts for this association in terms of the differing attitudes towards sexuality among different groups. She also finds that heaviness of menstrual flow, with the women's concurrent fear of staining, can account for most of the variance in affect.

Koeske finds in her studies that "... negative behavior exhibited premenstrually is perceived as evidence for the prevailing negative stereotype of female emotional behavior while positive behavior is ignored as something for which biology is irrelevant"(18; p. 12). Thus, premenstrual tension provides a social explanation for female "out of role" behavior, and the premenstrual display of such behavior confirms stereotypic views of female personality as changeable and unreasonable. Hostile behavior is deemed out of character for the female. Koeske's experiment on male and female students supported her hypothesis. The students judged biology more important for explaining premenstrual negative behavior than premenstrual pleasant behavior. She suggests, "if it can be demonstrated that premenstruation like epinephrine in Schachter and Singer's study is not sufficient to produce emotion, it would do much to challenge the negative images of women implicit in the biological explanation of premenstrual emotional behavior; it would make women look to their situation for explanations and change it perhaps, rather than blame themselves and their biology"(18; p. 13).

The work of Schraeder, Wilconan, and Sherif on 11 males, 11 females taking oral contraceptives, and 11 females not taking oral contraceptives (who filled out daily self-reports on pleasant activities, stressful events, moods, and somatic changes for 35 consecutive days) indicated that "stressful events accounted for more of the variance than did cycle phase for the negative mood factors but not for pain and water retention. There were large individual differences and heterogeneous variances"(19).

Alice Rossi, studying the interaction of biological and sociological factors in an attempt to bring biology back into sociology, and thus focusing on cyclicity in both females and males, found that "... a larger proportion of the total number of rated days included 'achy,' 'sick' and 'crampy' ratings among my *male* subjects than among my female subjects, while women recorded being 'happy,' 'friendly,' 'calm' and 'loving' for a larger proportion of the days they rated than men did"(4). Note that she studied positive as well as negative affect, in contrast to other researchers. From her analysis of two samples of students she studied she found(4):

1. *There is a firmer patterning of mood by the social week than by the menstrual month: many subjects showed no patterned shift of mood by menstrual cycle, but most showed such a pattern by calendar week.*
2. *Rather than a "blue Monday," the data suggested a "blue Wednesday," the day of the week furthest [sic] from the weekend, when mood states were most positive.*
3. *There was only a slight tendency toward a peak of positive moods at the point of ovulation, and a low point of negative moods the first few days of the menstrual period, but this was not a strong pattern, largely because so many subjects showed no patterning by menstrual cycle.*
4. *Where the two cycles were synchronized, e.g., if the first day of the menstrual period came on a Wednesday, or ovulation on a weekend, then the negative and positive peaking tended to show a sharp pattern indeed.*

Rossi also found that occurrence of intercourse did not fluctuate with the menstrual cycle, as some of the literature suggested it would because of supposed hormonally caused changes in sexual desire, but rather with the day of the week. It usually took place on weekends. Furthermore, women who were virgins were more likely to start menstruating on weekends than women who had sexual experience, presumably because this would give the former a legitimate (if subconscious) excuse to retain their virginity.

Alice Dan(20), studying cycles in eight married couples, found so much individual variability within phase and from one cycle to another and so much variability among her sample that she believes there are no consistent phase correlates. She notes that Dalton, whose work is frequently cited to support the position that women are not to be trusted during the premenstrual phase, only looks at negative effects. There may be positive effects, such as a lower threshold for sensations, that would account for such positive findings as a woman's reporting that her psychotherapy went better during the premenstrual phase.

Since Dan studied both women and men, she could compare them. If the menstrual cycle makes women's behavior more variable, then women should show more overall variance on the tests and self-reports, but, as a whole, men and women do not differ in variability, except for negative emotions. Women's feelings of anxiety and depression are more variable, but, according to the researcher, this can be tied to their oppressive situation rather than to a hormonal explanation, since the variations do not correspond to the cycle.

Parlee(5) has pointed out, in her review of the literature on premenstrual tension, that studies showing its absence are not published because journals tend not to publish negative results. Therefore, the body of literature attesting to its presence may be a result, not of its universality, but of the policy of the editors of learned journals. Furthermore, when she examined the primary sources about psychological fluctuations in mood and behavior through the cycle, she found less supporting evidence for many conclusions than is generally implied. When symptom check lists on premenstrual symptoms are administered to Swedish and Indian women, the results obtained in this country do not replicate. Moreover, according to Parlee, Joniger et al. stated in a review of cross-cultural data on premenstrual tension symptoms that "women in different cultures seem to report different clusters of 'symptoms' as the nucleus of a 'premenstrual tension syndrome.'"

For reasons of space, I will not discuss hormonal explanations for postpartum changes, except again to note that a focus on hormones takes attention away from the historically and cross-culturally unique situation in which young American mothers find themselves. After childbirth, they are totally isolated from female kin or supports, in many cases, with sole responsibility for the baby and no preparation for the life-style changes motherhood will bring about.

Parlee concludes her discussion of the literature about pregnancy, birth, and the postpartum period with a statement most appropriate for the issues we are discussing. She notes that the research in this area has neglected changes in social role occurring with childbirth, particularly negative aspects of parental roles, and ends, "By neglecting to attend to the sometimes negative consequences of traditional roles and role changes, and by focusing upon the individual woman in isolation from her social environment, scientists have, as in the study of the menstrual cycle, tended to emphasize those aspects of their data which, in the main, do not challenge the existing social order"(5).

There is little psychological literature on menopause compared with other life-cycle stages and events. Much of the clinical literature is hostile. Thus, in a National Institutes of Health conference on menopause, uncontaminated by the presence of a female although there were 25 male participants, one ob-gyn characterized menopausal women as being "a caricature of their younger selves at their emotional worst"(21; p. 3).

The diagnosis "involutional psychosis" applied to depressed women between the ages of about 45 and 55 years implies that their depression stems from the involution of the ovaries. However, my study of depressed middle-aged women(22) showed that it was loss of roles, not loss of hormones, that caused their problems. They had hormonal replacement therapy and that did not cause the depression to lift. The women had existential depressions resulting from the loss of their maternal roles, the roles that gave them their identity. They were supermothers and superwives, and they thought there would be a pay-off. Since there is no justice, there is no pay-off. They were considered useless, and they felt useless.

The cross-cultural study I undertook of menopause in 30 cultures showed that menopause did not seem to be a particularly difficult stage in the life cycle. Only those cultures investigated by women anthropologists had data on the menopause. But there were always data on the woman's position in a given culture and in most cultures her status improved. Since a rise in status affording one more privileges, more power, and fewer restrictions is likely to be associated with feelings of relative well-being, we can infer that women whose status increased had fewer problems at this stage in the life cycle than they had in earlier stages.

Doctors do more than just misdiagnose women's complaints. For example, in the "Biological Fallacies of Women's Lib," written by a physician (Gadpaille [23]) and published in a journal that was sent free to every M.D., we are treated to a particularly sloppy brand of biological determinism. The tone of the article was set by a picture on its cover of a woman naked to her breasts, save for a women's symbol with fist inside around her neck. He presents five so-called fallacies of women's liberation:

1. Brains—Male brains get androgen; therefore, they are organized differently. This difference affects animal behavior.
2. Genes—There are sex-linked diseases; thus, males are more fragile.
3. Hormones—"Women's lib accepts the facts of hormonal differences, too, but denies that this ubiquitous biological distinction affects behavior. Most practicing physicians know better"(23; pp. 38-39), for example, premenstrual tension causes women to have accidents more frequently just prior to menstruation.
4. Bodies—Men are taller, with heavier bones and muscles. Therefore, in the process of human evolution, natural selection has selected out males who are aggressive and physically powerful, since they are inclined to protect wives and offspring [or to beat them?] as well as females well suited to infant care and nurture and meeting needs of aggressive outer-directed mates.
5. Learning—The cortex (in chimpanzees) is not required for the female to have intercourse since she does not have to be conscious. Gadpaille then generalizes to the human female. "At the simple reproductive level, little or nothing is required of the female; conception can and does occur in states of violent opposition, psychological revulsion, and unconsciousness. For the male, sexual intercourse is contingent upon favorable psychological conditions permitting him to achieve an erection and maintain it long enough for intromission and ejaculation"(23; pp. 40 and 72).

Gadpaille uses the Money and Ehrhardt study of cross-sex behavior of fetally androgynized females(24) to support the importance of biological factors and then interprets the data politically. Male *domination* can be understood in psychodynamically logical terms because of the male's need to deny unconscious awareness of his greater fragility and vulnerability, both physical and psychosexual. In other words, they need to be aggressive because they are so weak—poor things. They used to be superior because they *were* superior. Now, it's because they are inferior—heads men win, tails women lose.

Let us see what, in fact, Money and Ehrhardt say about sexual dimorphism. They note the mutability of chromosomal sex in their studies of sexual anomalies. If an individual is by his or her chromosomes of one sex but is born with ambiguous external genitalia due to hormonal or other problems and is reared in the opposite sex, then the child's gender identity is the sex that she or he was *reared* in, even if it is opposite his or her chromosomal sex. If sex-change surgery is to be successful, it must occur before the child is 3 years old, because the gender identity is learned, with language, between 18 months and 3 years. In fact, such early changes have been successful. For example, a 7-month-old male's penis was destroyed in a circumcision. With the advice of the Johns Hopkins Clinic, he was then reared as a girl, with a superfeminine script followed by both parents. The child behaved in very traditional ways in her choice of toys, concern with cleanliness, and so on, although she did have lots of physical energy, which is considered male. As well as undergoing sex-change surgery, she will be given hormones later to develop secondary female characteristics such as breasts.

The Money and Ehrhardt research that is most often quoted to support ideas of biological determinism is concerned with fetally androgynized females, those females whose mothers were given male hormones while pregnant to prevent miscarriage. These girls tend to have higher IQs, wear slacks more, like dolls less, and generally engage in more tomboylike behavior. But the control group is inadequate. They should be compared, as endocrinologist Neena Schwartz points out, to other children who have had intensive attention from adults similar to that which these girls received at the Hopkins Clinic.[1] Moreover, we can all see the culture-based limitations of indexes such as wearing slacks and playing with dolls. It is important to realize that Money is no biological determinist, although his work is frequently cited to support a deterministic position. At most, he says, there are different *thresholds* for various behaviors, so that females have a lower threshold for maternal behavior than men, but human and animal males can care for infants.

In sum, Money and Ehrhardt's ovular (a term preferable to seminal) work used frequently by biological determinists, in fact, lends strong support to the mutability of biology.

Neuroanatomist Ruth Bleier(25) has made a number of important points while addressing herself to the issue of the relationship between anatomy and destiny. She notes that, while theories about nonhuman species and about the physical universe do not influence their behavior, theories about humans, such as theories about sex differences, are known by the subject and thus can change human behavior and function as self-fulfilling prophecies. Those of us who existed (I won't say lived) through the fifties can testify to the truth of that statement.

Next, she criticizes on three grounds the efforts of people, such as Ardrey(26-28) and Lorenz(29-33), who analogize from animal behavior to human behavior:

1. Use of selective data.
2. Ignoring culture, which makes humans different from other animals.
3. The fallacy involved when the ethologist interprets animal behaviors "using such terms as aggressive, submissive, amicable, dominant, etc., because of what they know about *themselves*—their own feelings, states, gestures and motivations"(25).

For example, they say that males are more aggressive because male rats fight more than females in a cage. This aggression is caused by androgen in the developing brain. In order to test this, newborn female rats were given androgen and found to fight more as adults. A later but less-publicized discovery was that newborn female rats given *estrogen* also fight more as adults.

Bleier continues with another example of the selective use of data with which investigators fail to interpret animal studies so as to contradict sex-role stereotypes. Although, in open-field tests, female rats are more exploratory than males and, in

[1]Personal communication with Neena Schwartz.

shock-avoidance tests, female rats learn faster than males, it is not concluded that female rats, or humans, are more intelligent than males.

Perhaps the most blatant use of selective animal data to reinforce traditional sex-role stereotypes occurs, not with rats, but with nonhuman primates. There, pop ethologists who seek justification for the sexual *status quo* love to write about male dominance among the rhesus macaque and the plains-dwelling African baboon. Both creatures are monkeys and are more distantly related to humans than are the great apes, among whom there is no clear relationship between sexual dimorphism and stereotyped sex-role behavior according to Leibowitz(34). Moreover, it has become clear that ecological factors greatly affect the behavior of nonhuman primates. Specifically, it has been found that monkeys and apes of the same species living under different ecological conditions behave quite differently with respect to sex roles. A prime example involves the baboon. The much-cited studies by the male supremacists involve baboons of the species *Papio anubis* living in the open plains. Yet, other *Papio anubis* African baboons that dwell in a *forest* environment do not exhibit the militaristic, hierarchically organized, male-dominant troops of their counterparts who live under high predation pressure on the broad savannahs (see Blumberg [35] for a review of the sex role implications of recent studies of nonhuman primates).

If discussions of the primacy of biological factors in human behavior, specifically in explanations of sex differences in behavior, were confined to academic journals, one might be less concerned about their political impact. But this position has been popularized in best-selling books written for lay audiences.

A popular academic presentation of a biological explanation of sex differences is Tiger's *Men in Groups,* in which Tiger presented the concept of male bonding, "the spinal column of a community"(36; p. 60), which he considers a human biogenetic inheritance. From this stance, he attributes women's lower status to their inability to create lasting all-female groups or to combine effectively in nonsexual relations with men. Thus, Tiger says, ". . . it may constitute a revolutionary and perhaps hazardous social change with numerous latent consequences should women ever enter politics in great numbers"(36; p. 205).

There have been numerous criticisms of his work from anthropologist Fried's fine review in *Science*(37) to psychologist Pleck's "Man to Man, Is Brotherhood Possible?"(38). Anyone looking at the Nixon administration, which was permeated to the core with machismo, has learned how soon the bonding leads to leaving the former teammate to twist slowly in the wind. (Sounds more like James Bonding than male bonding.)

As far as the absence of female bonding, historian Carroll Smith-Rosenberg(39) has meticulously documented the importance of female friendship in the nineteenth century, and a study by Seiden and myself(40) shows that most women have always had very close female friends and that the belief that women do not like each other is not true but, as an ideology, it performs a very important function. If one believes that women cannot be trusted, that women cannot be depended upon, and that women must of necessity compete with one another, there is only one other group to turn to—men. And that belief maintains the *status quo*.

THE KIBBUTZ

Let us now see how Tiger, who believes that sexual equality is biologically impossible, deals with the retreat from equality on the kibbutz and contrast his views with those of Mednick and Blumberg.

The kibbutz is an interesting case study with which to compare different models that explain the traditional sex-role allocation in a community that was supposed to say "Goodbye to All That"(41). All the studies agree on the fact that there is increasing familism on the kibbutz accompanying a very traditional allocation of work: women doing the child care, laundry, kitchen work, and teaching and men in the fields (the "productive" work). But the studies differ in their explanations. Lionel Tiger and Joseph Shepher, an Israeli sociologist and a student of Tiger, attribute the failure of the experiment to the inability of laws and attitudes to change people so that they will act so as to contradict, contravene, or distort "what may be natural mammalian patterns"(3; p. 13). The authors are trying to continue the trajectory that Tiger has been following about the relationship between biological *givens* and social *options.* They consider biology and sociology intrinsically related, but it is clear that the bedrock in their thinking is biology. For example, they ask the question, "Is sexual division of labor, in the widest sense, then a precondition of a successful mating system? In the widest sense, sexual division of labor means . . . that in a given population, the majority of men and women will not, over a period of time, be very responsive to ideologies or to technical and social changes that could blur that sexual polarization" (3; p. 240). Next, they note Darwin's decisive contribution of the "real link between sexual selection, natural selection and the consistent behavior of the species" after they present some data on the unattractiveness of dominant women to dominant men(3; pp. 240-41). They cite Abernathy's primatological, ethnographic, and psychiatric data that suggest that male dominance facilitates human male-female copulatory behavior (like rape?), whereas female dominance inhibits it(3; p. 241).

Like many biological determinists, they go directly from other animals to humans; for example, "We cannot forget the relevance of genetic transmission of behavioral characteristics. A chicken does not behave like a turtle or a cobra; a rhesus does not behave like an orangutan. These differences between them are anything but random. So it is with humans"(3; p. 271). Thus, they assume that because the genetic inheritance of turtles is different from that of cobras, resulting in different behavior, human behavior is similarly bound by a genetic code, and, thus, presumably, since males have an X and a Y chromosome and females two X chromsomes, differences in their behavior are immutable.

They explain the desire of kibbutz women to devote more time and energy to private maternal activities as follows: "Our biogrammatical assertion is that the behavior of these mothers is ethologically probable; they are seeking an association with their own offspring, which reflects a species-wide attraction between mothers and their young"(3; p. 272). They cite Bowlby (whose work on attachment has been questioned) on the necessity of mother-child interaction for the child's health and suggest that "both mothers and children in the kibbutz may be reacting to a strenuous violation of the biogrammar governing their relationship"(3; p. 274).

Next, they deliver their political sermon on the errors of feminists, although they admit the feminist movement is morally valid. They feel this moral validity prevents people from examining it unsentimentally. They believe feminist theorists to be wrong in their estimate of what women want; thus, necessarily, feminism fails to attract broad support. Tiger and Shepher believe that feminist theories will induce(3; p. 176)

> *. . . some of their followers to devote themselves so firmly to predominantly male patterns of work, politics, reproduction [reproduction?—how can a women devote herself to a* male *pattern of reproduction?] and values that they will* forgo an aspect of life they could enjoy along with others. *It is an irony that just when countless men and women and even nuns and priests are openly having liberal sexual relations, a new morality urges women to devalue and limit femininity and maternity. [My emphasis.]*

There are a number of errors here. First, one of the major points of a feminist analysis is that one cannot simply add traditional maternity to everything else one wants, given our social institutions. Second, they seem to confuse sexuality, or "liberal sexual relations," with "femininity and maternity." Have they never heard of contraception? Besides, few feminists are against voluntary maternity.

I will not point out the various errors in Tiger and Shepher's book, but rather I will present alternative models to explain what happened to the kibbutz. I should point out that theirs is a sophisticated book. They qualify their statements. They present a great deal of data and a relatively sophisticated statistical analysis. Their style is earnest. So, the book may convince people because of its apparent reasonableness and scholarship.

Before I discuss Mednick's and Blumberg's alternative models, let me point out that the kibbutz was never committed to real sexual equality. What it was committed to was the opportunity for women to do men's work but not for men to do women's work, like other socialist societies where women were free to dig ditches but men were not expected to wash dishes. Secondly, all labor was not equal. Productive labor, the labor that brought money into the kibbutz, was more highly valued than service labor. Since the former is traditional male work and the latter traditional female work, you can already see the seeds of the changes that took place.

Mednick(42) notes that most accounts of changes in the kibbutz use biological explanations. These are her reasons: first, stress on military preparedness, which made male values more important; second, devotion to physical labor and extreme emphasis on productivity and economic growth; third, affluence, bringing children, who increased the need for services and the demand for services so that, because it seemed natural for women to return to their traditional tasks, "the rationalization about women's true nature was reactivated as a way of reducing the anxiety raised by the discrepancy between the ideology of equality and emancipation and the reality of the behavior in which women were engaged"(41; p. 97); and, fourth, pronatalism fed by nationalism and the communities' own motive to survive—called internal immigration (supplementing immigration from other countries). Also, having children is the most valued nonmasculine kibbutz role, so it may be the kibbutz woman's way of being a productive kibbutz member.

Rae Blumberg(7; p. 319) notes that most of the literature on the kibbutz presents the position that the drift of women back to their traditional roles—from liberation to laundry—is explained primarily because of the "psychological and/or biological make-up" of the women themselves, a blaming the victim approach. It is alleged that they drift into service occupations for physical, psychological, and "natural" reasons. These explanations emphasize factors that *pull* women into domestic service and child-care jobs. She proposes a *push* explanation and emphasizes that women were shoved into service activities and *de facto* second-class citizenship. Blumberg does not agree that laundry is destiny. Rather, she proposes that the initially favorable position of kibbutz women was eroded by a series of demographic and structural constraints and contradictions. The latter emerged from the way the kibbutzim operationalized their deliberately chosen mode of production, agrarian socialism. For example, one factor with negative implications for women's fate springs from the kibbutz's socialist tenet that "productive" (capital-generating) labor is more highly valued than capital-consuming services. As we will see, men were always more likely to engage in the former and women gradually became concentrated in the latter.

Originally, in the early pioneering phase, the kibbutz was characterized by substantial (although not total) sexual equality. Women (only 20-35 percent of the kibbutz founders) worked side by side with the men in the fields and participated actively in the governance of the kibbutz; the service sector was minimal (although women got stuck with a disproportionate share even then), and births were rare. Currently, there exists a highly sex-differentiated division of labor: men involved in highly esteemed productive and managerial work and women engaged in the service activities from which they ostensibly had been liberated. Concomitantly, women are underrepresented in both kibbutz offices and major committees that control the kibbutz political economy. Research has shown that women holding the less-skilled and low-prestige service occupations (e.g., laundry) are more likely to want more private family responsibilities, such as cooking some meals for their families and having children sleep in their apartments; while, conversely, those women holding office or working in highly valued (agricultural) activities are less likely to want such changes. Unless we assume biological differences between these two categories of women, it appears that it is a women's social position and her occupational dissatisfaction, rather than her biograms or hormones, that drive her to "Kind, Kirche, Küchen."

In the physical and psychological reasons most authors use to explain the women's drift into service occupations, women are considered the main initiators of this response. In contrast, Blumberg's *push* explanation focuses on the one-way integration of women into "male" roles with little integration of men into "female domestic roles" (apparently, men never were assigned to the nurseries, for example). Women's allocation to the service sector (kitchen, laundry, child care, etc.) was augmented by the arrival of children and immigrants. The former precipitated an increase in the size of the service sector and the latter—who, like the initial kibbutz pioneers, were predominantly single, childless, and male—provided an alternate labor force for the agricultural sector that made it possible to *replace* those young kibbutz mothers in the fields.

According to Blumberg, contradictions in the kibbutz's mode of production have eroded the position of women in a number of ways. As noted, the kibbutz adopted agrarian socialism as its mode of production. "Agrarian" refers to their techno-economic base, which revolves around plow cultivation of cereal crops; unfortunately, this base throughout history has always been associated with low female participation in production and with sexual inequality. "Socialism" refers to their collectivized social relations of production and the associated ideology; unfortunately, they never worked out the negative implications of some of their practices on their ideological tenet of sexual equality. Among the sexual-equality-eroding contradictions:

1. The rational economy of the kibbutz, which strives to maximize return within the constraints of its ideology. This approach favored agrarian dry farming and large-animal husbandry as opposed to horticultural and small-animal activities, which generally have a predominantly female labor force. In fact, the kibbutz bookkeeping system, based on Marx's labor theory of value, further exacerbated the trend toward male-oriented agrarian production. The kibbutz early adopted a bookkeeping yardstick, "income per labor day," that calculates only the amount of labor put into an activity (i.e., ignores land and capital costs). Plow agriculture and large-animal husbandry use more land and less labor per unit of output than the horticulture and small-animal tending that are preferred by the kibbutz women and so look better in the kibbutz ledgers. Over the years, the productive activities favored by the women have gradually been phased out for economically "rational" reasons.

2. The role of immigrants. The kibbutz, as a socialist community founded in hopes of being a revolutionary vanguard, was committed to both expansion and the principle of "self-labor," that is, not employing hired workers from outside the kibbutz. Accordingly, the kibbutz was committed to attracting and keeping immigrants. And, during the early years of the kibbutz, a steady stream arrived. The

majority were single males. By this time, the kibbutz pioneer women had begun having babies, increasing the demand for service activities. Combining responsibilities to children with work in the distant fields (see the next point) was very inconvenient for the kibbutz pioneer mothers. Clearly, the unencumbered immigrants represented a preferable labor force for the kibbutz's "glamour" sector, the agrarian field crops. And, had they been relegated to the much-disliked service tasks, they might not have stayed, while the women were tied to the kibbutz by their children, husbands, and years of dedicated toil.

3. Kibbutz ecology versus child psychology. The prevailing child psychology of the kibbutz was not specifically socialist; in fact, it emphasized the importance of breast feeding and mother-infant contact. This belief system meant that mothers had to work near the children, not far away in the fields. The mothers could have worked near the children had horticultural activities been a greater component of kibbutz production, since these tend to be located physically closer to the center of the kibbutz and the children's housing.

4. Growth of the service sector and the accelerating downward spiral. Women began to be concentrated in the low-valued service tasks just as an alternate labor force permitting their replacement in the fields became available. Meanwhile, the kibbutzim turned the corner from adversity to greater affluence. This was reflected in more babies and higher living standards. Both factors, combined with the availability of females no longer needed in the fields, ballooned the service sector and froze females into its expanding sphere. As women left the fields, they held fewer places on the crucial economic committees. So too their political participation. But it was not a contented retirement into the comforting confines of women's traditional roles: a number of empirical studies have found considerable occupational dissatisfaction among women stuck with the "domestic drudgework" of service jobs. And it was such women, those most unhappy in their "nonproductive," low-valued jobs, who were most likely to push for more private family responsibilities. In recent years, large proportions of men also have pressed for greater individual familism. Yet, no one has proposed male biograms as the underlying cause.

CONCLUSION

We are seeing a resurgence of interest in biological determinants of behavior. While in itself such an interest is as legitimate as any other intellectual interest, we know that, in certain areas of human experience and behavior, "knowledge" has political consequences. We know that biological explanations of human behavior have traditionally been used to legitimize the *status quo*—racism, sexism, and the existing class structure. This paper specifically dealt with biological determinism and sexism. There is nothing intrinsically oppressive in the examination of the impact of biological variables on human experiences. However, white women, as well as non-white men and women, must be suspicious of such interest because of the past use of biological determinism to legitimize sexism and racism. Moreover, most of the literature suggests that biology affects women more than men, the latter presumably freer to be creatures of reason rather than subject to the "raging hormonal influences" said to characterize women. And it is probably no accident that the popularity of books dealing with biological factors in human behavior coincides with a decade reminiscent of the fifties in conservatism and lack of student activism.

We know there is, to use Max Weber's phrase, an "elective affinity" between theories of biological determinism and conservative ideologies because of the assumption that what is biological cannot be changed. It is important that biological arguments used against feminism be critically examined and, when found wanting, refuted. It was this task to which this paper was addressed.[2]

References

1. Short, R. Z. 1972. Sex determination differentiation. In *Embryonic and fetal development,* eds. C. R. Austin and R. V. Short. Cambridge: Cambridge University Press.
2. Tarvis, C. 1973. Who likes women's liberation—And why: The case of the unliberated liberals. *The Journal of Social Issues* 29(4):175-98.
3. Tiger, L., and Shepher, J. 1975. *Women in the kibbutz.* New York: Harcourt Brace Jovanovich.
4. Rossi, A. S. 1974. Physiological and social rhythms: The study of human cyclicity. A paper presented at the American Psychiatric Association, Detroit.
5. Parlee, M. 1975. Aspects of menstruation, childbirth, and menopause: An overview and suggestions for further research. A paper presented at the Conference on New Directions for Research on Women, Madison, Wis.
6. Lennane, J. K., and Lennane, J. R. 1973. Alleged psychogenic disorders in women—A possible manifestation of sexual prejudice. *New England Journal of Medicine* 288(6):288-92.
7. Blumberg, R. 1976. Kibbutz women: From the fields of revolution to the laundries of discontent. In *Women in the world: A comparative study,* eds. Lynne Iglitzin and Ruth Ross. Santa Barbara, Calif.: ABC Clio.
8. Barker-Benfield, G. J. 1976. *The horrors of the half-known life.* New York: Harper and Row.
9. Haller, J. S., and Haller, R. M. 1974. *The physician and sexuality in Victorian America.* Urbana, Ill.: University of Illinois Press.
10. Coleman, R. 1889. Woman's relations to the higher education and professions, as viewed from physiological and other standpoints. Medical Association of the State of Alabama, *Transactions.* [Quoted in J. S. Haller and R. M. Haller.·1974. *The physician and sexuality in Victorian America.* Urbana, Ill.: University of Illinois Press.]
11. Hutchinson, W. 1895. The economics of prostitution. *American Medico-Surgical Bulletin* V. [Quoted in J. S. Haller and R. M. Haller. 1974. *The physician and sexuality in Victorian America.* Urbana, Ill.: University of Illinois Press.]
12. Colombat, D. 1845. *Colombat on the diseases of females: A treatise on the diseases and special hygiene of females.* Trans. and ed. Charles D. Meigs. Philadelphia: Lea and Blanchard. [Quoted in D. Scully and P. Bart. 1972. A funny thing happened on the way to the orifice. A paper presented at the American Sociological Association, New Orleans.]
13. Cooke, W. R. 1943. *Essentials of gynecology.* Philadelphia: Lippincott.
14. Parsons, L., and Sommers, S. C. 1962. *Gynecology.* Philadelphia: Saunders.
15. Green, T. 1971. *Gynecology: Essentials of clinical practice.* Boston: Little, Brown and Company.
16. *The New York Times.* July 26, 1970, p. 35.
17. Paige, K. E. 1973. Women learn to sing the menstrual blues. *Psychology Today* 7(4):41-46.
18. Koeske, R. D. 1976. Premenstrual emotionality: Is biology destiny? *Women and Health* 1(3):11-14.
19. Schraeder, S. L., Wilconan, L. A., and Sherif, C. W. 1975. A new psychology of menstruation. A paper presented at the American Psychological Association, Chicago.
20. Dan, A. 1976. Patterns of behavioral and mood variation in men and women. Dissertation at the University of Chicago, Committee on Human Development.
21. Jones, N. W., Jr., Cohen, E. J., and Wison, R. B. 1971. Clinical aspects of the menopause. In *Menopause and aging,* eds. J. K. Ryan and D. C. Gibson. U.S. Department of Health, Education, and Welfare publication no. NIH 73-319. Washington, D.C.: U.S. Government Printing Office.
22. Bart, P. B. 1971. Depression in middle-aged women. In *Women in sexist society,* eds. Vivian Cornick and Barbara K. Moran. New York: Basic Books.
23. Gadpaille, W. 1971. The biological fallacies of women's lib. *The Hospital Physician,* July 1971, pp. 36-40, 72-74.
24. Money, J., and Ehrhardt, A. S. 1972. *Man and woman, boy and girl.* Baltimore: Johns Hopkins University Press.
25. Bleier, R. 1973. Is anatomy destiny—The second X. A paper presented at the American Sociological Association, New York City.

[2]I wish to thank Marlene R. Drescher for general assistance in the preparation of this paper and Rae L. Blumberg for bringing to my attention the work on primates.

26. Ardrey, R. 1961. *African genesis.* New York: Atheneum.
27. Ardrey, R. 1966. *The territorial imperative.* New York: Atheneum.
28. Ardrey, R. 1970. *The social contract.* New York: Atheneum.
29. Lorenz, K. 1952. *King Solomon's ring.* New York: Crowell.
30. Lorenz, K. 1955. *Man meets dog.* Boston: Houghton Mifflin Company.
31. Lorenz, K. 1966. *On aggression.* New York: Harcourt Brace Jovanovich.
32. Lorenz, K. 1970. *Studies in human behavior.* Cambridge, Mass.: Harvard University Press.
33. Lorenz, K. 1973. *Motivation of human and animal behavior.* New York: Van Nostrand.
34. Leibowitz, L. 1975. The evolution of primate sex differences. In *Toward an anthropology of women,* ed. Rayna Reiter. New York: Monthly Review Press.
35. Blumberg, R. L. 1976. *Stratification: Socio-economic and sexual inequality.* Dubuque, Iowa: William O. Brown.
36. Tiger, L. 1969. *Men in groups.* New York: Random House.
37. Fried, M. H. 1969. Mankind excluding woman. *Science* 165(3896):883-84.
38. Pleck, J. H. 1975. Man to man, is brotherhood possible? In *Old family/New family,* ed. Nona Glazer-Malbin. New York: Van Nostrand.
39. Smith-Rosenberg, C. 1975. The female world of love and ritual: Relations between women in nineteenth-century America. *Signs: Journal of Women in Culture and Society* 1(1):1-29.
40. Seiden, A. M., and Bart, P. B. 1975. Woman to woman: Is sisterhood powerful? In *Old family/New family,* ed. Nona Glazer-Malbin. New York: Van Nostrand.
41. Morgan, R. 1970. Goodbye to all that. *Rat,* February 7, 1970.
42. Mednick, M. T. S. 1975. Social change and sex-role inertia: The case of the kibbutz. In *Woman and achievement,* eds. M. T. S. Mednick, S. S. Tangri, and L. W. Hoffman. New York: Halsted Press.

Aggression

During the 1960s, the violence in cities, on college campuses, and in Vietnam resulted in increased concern over the causes of aggressive behavior. The opinion that human beings are innately aggressive became very popular as a result of the publication of Konrad Lorenz's *On Aggression*(1) and Robert Ardrey's *The Territorial Imperative*(2) and *African Genesis*(3). Lorenz and Ardrey tried to show that the roots of human violence lie in our genetic heritage.

This hypothesis gained credence in 1965, when it was discovered that a number of males in prisons had a chromosomal abnormality, an extra Y chromosome (normal males have one X chromosome and one Y chromosome; these males had one X and two Ys). The XYY males were found in prisons in higher proportions than would be predicted by the frequency with which this abnormality occurs in sex-cell formation. The "obvious" conclusion was that this extra Y chromosome was the cause of their criminal behavior. Since it was (and still is) assumed that males are by nature more aggressive than females and since females do not have the Y chromosome, it seemed natural to conclude that the Y chromosome carries genes for aggressive behavior. A group of scientists in Boston doubted these conclusions. In the first paper in this section, some of those scientists tell why they rejected the deterministic view of aggression in XYY males, and they describe the problems created by such studies.

In the second paper, Richard Kunnes asks why violence is viewed as something done only by individuals. Are not corporations and governments violent as well? If so, how is it possible to promote a genetic basis for violent behavior? Kunnes sees violence as a product of our socioeconomic system and discounts the genetic hypothesis as a misleading view that functions to distract us from promoting social change.

The authors of both papers seek to dispel the notion that aggressive behavior is natural and inevitable.

References

1. Lorenz, K. 1966. *On aggression.* New York: Harcourt Brace Jovanovich.
2. Ardrey, R. 1966. *The territorial imperative.* New York: Atheneum.
3. Ardrey, R. 1961. *African genesis.* New York: Atheneum.

The XYY Male: The Making of a Myth

Reed Pyeritz
Herb Schreier
Chuck Madansky
Larry Miller
Jon Beckwith

An area of human behavior that has consistently absorbed the interest of many scientists is the study of the origins of criminal, antisocial, and aggressive behavior.[1] In the late nineteenth century, the Italian criminologist Cesare Lombroso proposed that there was a correlation between physical type and criminal behavior(6,7). His theories, which achieved wide circulation and popularity, included rather specific suggestions to law enforcement authorities for detecting criminals: according to Lombroso (via Ellis): "Thieves have shifty eyes, bushy eyebrows, receding foreheads and projecting ears. . . . Murderers have cold, glassy bloodshot eyes, curly abundant hair, strong jaws, long ears, and thin lips"(8; p. 32). Or, more generally, as Haller puts it: "The criminal tended to have a primitive brain, an unusual cephalic index, long arms, prehensile feet, a scanty beard but hairy body, large incisors, flattened nose, furtive eyes and an angular skull"(2; p. 15). This approach to "criminal anthropology" was later adopted by Ernest Hooton, an anthropologist at Harvard, who argued for the "criminal career based on the racial heredity of the individual"(1; p. 179). Hooton again claimed a connection between criminal acts and physical features(9; p. 307).

With the emergence of modern genetics at the turn of the twentieth century many scientists turned to more explicitly hereditarian arguments to explain criminal behavior. While Francis Galton, the founder of eugenics, had made such proposals during the nineteenth century, the rediscovery of Mendel's laws in 1900 appeared to add more weight to these theories. Claiming the backing of genetic knowledge, Harry Laughlin of the Eugenics Record Office at Cold Spring Harbor, New York, proposed that a significant proportion of the population of the United States was genetically defective, and so included individuals who were "feebleminded, insane, criminalistic," and so on(2; p. 133). Lewis Terman, involved in introducing the Binet intelligence test into the United States, firmly believed in the genetic basis of IQ and argued that "all feebleminded are at least potential criminals"(10; p. 11). The prestigious *Harvard Law Review* added its weight to the genetic arguments by stating that "larceny is common among born criminals . . . "(11; p. 123).

[1]The references cited as 1 through 5 in the reference list served as the sources for information presented in this paper.

The concept of hereditary criminality was further reinforced by the publication in 1912 and 1916 of two books that purported to show the genetic transmission of all sorts of socially undesirable traits in two notorious families—the Jukes and the Kallikaks(12, 13). Henry Goddard, another IQ tester, proposed in his study of the Kallikak family that Martin Kallikak had sired both a good and bad side to his family tree. Through his involvement with a feebleminded girl, he had contributed "paupers, criminals, prostitutes and drunkards," while the children born to his wife became "hard-working, law-abiding citizens."

Nearly all of the scientists cited above were members of the eugenics movement, which had a significant influence on social policy in the early twentieth century. The eugenicists lobbied successfully for state sterilization laws that were supposed to prevent the further contamination of the gene pool. The preamble to the first such law, passed in Indiana in 1907, stated, "Whereas heredity plays a most important part in the transmission of crime, idiocy and imbecility . . ."(3; pp. 10-11). Over the next several decades, a total of 33 states passed sterilization laws. One-half of these included provision for sterilization of sex criminals, and at least 6 states allowed sterilization for a much broader range of crimes. While most sterilizations performed under these laws were for feeblemindedness or insanity, there were also cases involving prisoners in penal institutions(14; p. 135).

As a result of the reaction against the Nazi extension of eugenics ideology and of the increased attention to social and economic determinants of antisocial behavior, the arguments for a biological basis for criminal behavior faded into the background in the forties and fifties. It was not until the late 1960s that the old arguments surfaced once again, cloaked in the language of the latest biological and medical advances. For instance, three doctors involved in psychosurgery experimentation wrote, in a letter to the *Journal of the American Medical Association,* in 1967(15; p. 895):

It is important to realize that only a small number of the millions of slum dwellers have taken part in the riots. . . . Is there something peculiar about the violent slum dweller that differentiates him from his peaceful neighbor?

There is evidence from several sources . . . that brain dysfunction related to a focal lesion plays a significant role in the violent and assaultive behavior of thoroughly studied patients. . . .

We need intensive research and clinical studies of the individuals committing the violence.

The genetic arguments were raised once again by H. J. Eysenck, who proposed in 1964 ". . . some kind of gene, chromosome, or other structure which could be the physical basis for differences between the criminal and non-criminal type of person"(16; p. 61). Remarkably, within a year of this suggestion, the increasing sophistication of genetic technology and, particularly, the ability to examine and distinguish human chromosomes had permitted a new "discovery"—"the criminal chromosome." What this phrase refers to is the belief, which was held by a number of scientists, that males born with an extra Y chromosome—XYY males—were doomed to a life of criminal behavior.

THE ORIGINS OF THE XYY MYTH[2]

The XYY chromosome variation was first noted in 1961, when a case was reported that had been discovered during screening for abnormal chromosomes(20). The next few years brought scattered case reports, but it was not until 1965 that XYY and

[2] Recent reviews of the XYY literature are cited as numbers 17, 18, and 19 in the reference list.

behavioral problems were linked. In that year, Patricia Jacobs and her coworkers reported that there was a significant proportion of XYY males among mentally subnormal inmates in a Scottish institution for those with "dangerous, violent or criminal propensities." This research was reported in an article with the provocative title, "Aggressive Behavior, Mental Subnormality, and the XYY Male"(21). A series of studies that followed also restricted screening to prisons, mental hospitals, and institutions for the mentally retarded and largely ignored screening of control populations. Since Jacobs et al. had reported that the XYY males were, on the average, taller than XY males, some of the researchers further limited their screening to only tall inmates. These studies resulted in claims even more provocative than those of the original report. One group of British researchers in 1967 found it "reasonable to suggest that antisocial behavior is due to an extra Y chromosome"(22; p. 815), and another group in Denmark in 1969 concluded that "patients with the XYY syndrome have pronounced disposition to personality deviation and criminality"(23; p. 965). Such mutually consistent "results" lent an air of scientific authenticity to the "XYY syndrome."

Despite the absence of any solid evidence for a relationship between XYY and antisocial behavior, the suggestive titles and conclusions of these studies brought wide publicity to the claims. The publicity, in turn, fostered the general impression among the public that such a relationship did indeed exist. While lurid articles appeared in the popular press, including a *Newsweek* article entitled "Congenital Criminals"(24), the XYY myth was also rapidly incorporated as fact into high school and college texts. For instance, a recent high school biology text included the following statement: "Another abnormal condition results when a normal X-bearing egg is fertilized by a YY sperm, formed by non-disjunction during spermatogenesis. This produces an XYY male who is usually over six feet in height and *very aggressive*"(25; p. 185). [Our emphasis.]

The XYY myth was further fortified in the United States when it was incorrectly reported that Richard Speck, murderer of eight student nurses in Chicago in 1968, was an XYY male. Despite repeated karyotypes that showed that Speck was an XY male(26), the retraction that appeared shortly after the article was published received little attention and the original claim is still generally believed. For example, a 1972 standard text on psychiatry included a picture of Speck with the following legend: "His [Speck's] legal defense was based on his genetic makeup which was 'XYY'. Individuals with these genes have been reported to be tall, be mentally retarded, have acne, and show aggressive behavior"(27; p. 711). There even appeared a popular novel entitled *The XYY Man*(28), the main character of which, an XYY individual, is just out of prison and is working as a burglar for British Intelligence.

Beyond the widespread acceptance of the XYY myth, there have been direct social consequences of the assumption that XYY males were destined for a life of antisocial behavior. In at least two states in the United States (Massachusetts and Maryland), adolescent males in institutions for juvenile delinquents have been screened for an extra Y chromosome. In Maryland, only a prolonged court battle prevented this information from being handed over to the juvenile courts(29), where, given the current climate of belief, it might have influenced sentencing. Furthermore, in certain institutions, XYY inmates were treated with female sex hormones in an attempt to restore "normal" behavior(30, 31).

Bentley Glass, former president of the American Association for the Advancement of Science, looked forward to the day when a combination of amniocentesis and abortion would "rid us of . . . sex deviants such as the XYY type"(32; p. 28). In fact, there is already at least one reported case of a mother opting for abortion after learning that her fetus was XYY and hearing from the doctor "what little was known of the prognosis at the time"(18; p. 150). Ashley Montagu, referring to the XYY male, suggested that we "must consider seriously . . . whether it would not now be desirable to type chromosomally all infants at birth or shortly after. . . . Forearmed with such information, it might be possible to institute the proper preventive and other measures at an early age"(33; p. 49).

However, 11 years after the original study and more than 200 XYY research articles later, the state of the field is such that the most recent and comprehensive review of this area concludes that "the frequency of antisocial behavior of the XYY male is probably not very different from non-XYY persons of similar background and social class"(17; p. 188). And, most recently, Ashley Montagu has corrected his earlier misimpressions: "Immediately all sorts of premature and unwarranted conclusions were drawn. . . . It is clear that an extra Y chromosome is not a 'violent' chromosome. . . . The history of the XYY anomaly constitutes an object-lesson in how not to draw conclusions about causation from conditions that happen to be associated"(34; p. 220).

How did we arrive at a situation in which such a well-supported field of research has become, for the most part, discredited? How did it come about that the false conclusions put forth by so many researchers had such an immediate and extensive social impact? In order to answer these questions, we must look at both the nature of the research itself and at the social and political milieu in which it developed. An inspection of the research reveals a host of basic defects: methodological errors, the use of subjective categories in defining aggression and antisocial behavior, and a failure to distinguish association and correlations from causation. In the next sections, we review these problems and then suggest how the atmosphere of the 1960s may have played a role in generating the XYY myth.

METHODOLOGICAL ERRORS

The methodology employed in the overwhelming majority of XYY studies has been seriously flawed(17, 18, and 19). In general, studies have lacked sufficient controls and conclusions have been based on scanty data. Five types of studies comprise the majority of XYY research done to date, each of which is subject to severe limitations. These are prevalence studies, incidence studies, retrospective studies, prospective studies, and case studies. We will examine each class of study, in turn, and then summarize what little information has come out of these studies.

Prevalence Studies

Prevalence studies identify the frequency with which the XYY genotype occurs in any given group at a particular time. Investigators have most often done such studies with only those groups that they thought might be most likely to contain XYY individuals. For example, the majority of prevalence studies have been conducted in institutional settings, a reflection of the initial studies that found a number of XYY males in mental-penal facilities. Other surveys have provided evidence that XYY males tend to be taller than their XY counterparts, leading some researchers to focus only on height as a selection criteria. Thus, "tall athletes and tall men who were fairly successful in their professional careers and who belonged to a special club" were screened(35). However, in few of these studies was there an appropriate control group, matched for height, race, socioeconomic class, institutional status, and so forth, that was screened for XYY as well. In the absence of an accurate demographic picture of the study population and adequate control groups, interpretation of the prevalence data for XYY males should have been severely limited. For example, there was no evidence provided by the vast majority of studies to date that could refute the suggestion that an increased frequency of XYY males in certain populations (e.g., institutionalized ones) was due to a statistical association between the extra Y chromosome and increased height! Yet, researchers used such evidence to claim further support for the supposed connection between XYY and antisocial behavior.

Incidence Studies

Incidence studies identify the rate of appearance of newborn XYY males. Incidence rates could differ from prevalence rates for the general population if there were a differential mortality between XY and XYY males. (No evidence for such a differential exists.) These studies suffer from a problem analogous to that of prevalence studies: the screening occurs in particular maternity hospitals whose clientele may not be representative of the general population. In one case, the data may have been further distorted when only non-private-ward babies were screened(36). In many instances, the sample has been too small to tell statistically what is the frequency of XYY babies. In many instances, the sample had been collected over too short a time and contained too few individuals to permit statistical validation of the frequency of XYY males. Investigators compounded these deficiencies when they pooled data from different sources (hospitals and countries). Studies from different sources have yielded different frequencies, and the factors accounting for these differences have not been defined.

Retrospective Studies

Retrospective studies seek first to identify XYY men in a defined population and then to examine their past history, through records or interviews, for details of interest, such as crimes committed. A major limitation of retrospective studies is that these populations are determined by the availability of records. For this reason, one major retrospective survey was carried out in Copenhagen, Denmark, where birth, education, and military records are maintained on everyone(37). However, in defining the "intelligence" of the survey population, the researchers had to rely on inadequate military induction tests because these were the only data available. Moreover, in this study, only tall men were karyotyped because it would have been too expensive to screen the entire male population. Restrictions on the design of a study such as these make it all the more difficult to extend the conclusions to the general population or even to all XYY males. While some of the recent retrospective studies have included control groups (for example, Witkin et al. [37] and Noel et al. [38]), most of the earlier ones (Zeuthen et al. [40], for example) did not. Thus, the information from the earlier studies is, at best, anecdotal. Compounding these errors is the problem that researchers carrying out such studies have often been aware who is and who is not an XYY male. For this reason, their interpretations of psychological tests and interviews and their analyses of even retrospective data may well have been biased.

This last criticism exposes a problem endemic to psychological studies: the failure to use double-blind procedures. The double-blind approach requires that there be among the subjects studied a control group as well as an experimental group of interest and that neither the researchers nor the subjects should be aware of who is in the control group and who is in the experimental group. (A third requirement, seldom met, is that neither the subjects nor the individuals who collect the data should know the *purpose* of the study.)

In his book *Introduction to Scientific Research,* Professor E. B. Wilson wrote(39; pp. 43-44):

Randomization can prevent human bias from entering in the selection of the sample and again in making the assignment of subjects and controls, but after this stage the experiment is still in grave peril from the ever-present danger of psychological distortion. . . . It is not merely the subjects who are influenced by psychological factors: the experimenter himself can easily be deceived in interpreting the results by his personal interest in the outcome. In the best experimental designs, the person making comparisons, measurements, or records is kept ignorant of the identity of the subjects and controls. Even in such routine matters as recording long lists of numbers or other single data, it has been demonstrated that the mistakes which are made are usually more numerous in the direction personally favored by the recorder. No human being is even approximately free from these subjective influences; the honest and enlightened investigator devises the experiment so that his own prejudices cannot influence the result. Only the naive or dishonest claim that their own objectivity is a sufficient safeguard.

The double-blind approach avoids biases of researchers in evaluting subjects and the effects of subject knowledge on their own behavior. Unfortunately, both retrospective and, as we shall see, prospective studies often fail to employ even a single-blind approach.

Prospective Studies

Prospective studies identify XYY individuals and then follow their development. Such studies are highly subjective in the kind of information the researcher compiles and in how he or she evaluates that information. This is true, for instance, in the case of following newborns when one relies on the behavioral descriptions of children by parents and the ability of the investigator to interpret the descriptions correctly.

But the major difficulty is that both requirements of the double-blind principle are usually violated. For example, in a study at Harvard Medical School, the researchers knew which children were XYY, and most families knew also. Given the widespread publicity about XYY and "criminality genes" and the investigators' offer of intervention "in case" any troubles should arise, the knowledge that their children were XYY was likely to raise parental anxieties. In the words of one of the researchers in this study, "We are fully aware that such information will influence the rearing of the children and thereby bias the results of the study"(36). The effect of this intervention in the family's life and of parental concern could well be the appearance of the very behavior feared or of other behavioral problems in the child(41).

Case Reports

Finally, there are numerous case reports in the literature. As little as one individual case of an XYY individual, complete with subjective psychological profiles, has been presented as evidence for a psychopathology associated with the XYY genotype. Since these individuals were noticed because they exhibited unusual behavior to begin with, they can hardly be considered representative of XYY males. One such case study is that described by Jarvik et al.(42). Their characterization of the particular XYY male as deviant relied, in part, on interviews with his mother. During one such interview the mother revealed that "he had a history of antisocial behavior dating back to early childhood when . . . she finally tied him to a tree when doing her outdoor chores in order to prevent him from harming himself and others"(42; p. 676). This is taken as evidence by the authors of the child's unusual behavior rather than as a possible indication of some psychopathology on the part of the mother. Could such treatment of the child have played a role in later behavioral patterns?

Despite the host of methodological errors in XYY studies, it is still possible to glean some tentative information from the work that has been done. XYY males are, on the average, taller than their XY counterparts and their own fathers. Further, in the *mental-penal* institutions (facilities for the criminally insane) that have been studied, the prevalence of XYY males tends to be higher than would be predicted on the basis of preliminary frequencies for the general population(18). However, even accepting the likelihood that these data are significant (matched control populations have not yet been examined for the XYY character), those in mental-penal settings represent a very small population. As a result, it can be estimated that only about 1 percent of XYY males are likely to be found in such institutions(43). This estimate is, in turn, based on the results of newborn studies that have given incidence frequencies of from 1/1400 to 1/500, depending on the hospital. There is now almost universal agreement that XYY males are not more likely to be found in *penal* institutions than XY males. Likewise, there is no direct evidence that XYY males are more likely to commit crimes. The vast majority of XYY males who have been detected appear to be leading socially normal lives. Apart from increased height and a slight risk of being confined in a *mental-penal* institution, no other characteristic—physical, mental, or social—has been satisfactorily documented to be associated with the extra Y chromosome.

DEFINITIONS OF AGGRESSION AND ANTISOCIAL BEHAVIOR

A second major class of defect in XYY research has been the subjective nature of the categories used to classify XYY individuals. In particular, the categories of "aggression" and "antisocial behavior" are by no means well defined. Early studies, such as that of Jacobs and her coworkers(21), sought to define aggression pragmatically; that is, those males incarcerated in mental-penal institutions were thought by definition to have committed aggressive acts. Other studies, particularly those involving XYYs discovered in the general population, relied on more obviously subjective criteria, such as observer judgments and psychological testing. Several investigators attempted a prior genetic explanation: aggression was linked to the Y chromosome, and it was assumed that those with an extra Y chromosome would be abnormally aggressive. All three attempts at categorizing aggression are seriously defective and cast doubt on the basic assumptions of much XYY research.

Those studies identifying aggression with the Y chromosome are the most obviously flawed because they rely on a dubious supposition: that males, having an XY chromosomal complement, were "known" to be more "aggressive" than females. If the Y chromosome had some relation to the increased aggressiveness, then an extra Y chromosome might confer even more aggressiveness. Jarvik et al. make this reasoning explicit: ". . . the Y chromosome is the male-determining chromosome; therefore, it should come as no surprise that an extra Y chromosome can produce an individual with heightened masculinity, evinced by characteristics such as unusual tallness, increased fertility, . . . and powerful aggressive tendencies"(42; pp. 679-80). This supposition is simplistic to the extreme. In addition, the investigators ignore the obvious contribution of different socialization processes for males and females.

The second type of study involved attempts to enumerate the behavior characteristics of XYY males. A tremendous range of behaviors was attributed to XYYs by various investigators. Included are such heterogeneous terms as "schizoid," "suspicious," "sociopathic," "loners," "immature," "excessive daydreaming," "irresponsible," "inadequate," "passive aggressive," and "hyperactive." In a 1972 review, Owen classified the behavioral descriptions used up to that time. The most commonly employed categories were immaturity, impulsivity, temper tantrums, and aggression(44). But, as Owen points out, this data is at best impressionistic: little effort was made to design "blind" procedures to avoid experimental bias, and there is no indication that characteristics are similarly defined by different investigators. Brief examination of several more recent studies illustrates these problems. One group compared five XYY males to their siblings and found the XYYs to be more immature, more impulsive, and more prone to "contact difficulties"(40). Some XYYs were described as "restless" or "hot-tempered." The evidence provided for these conclusions was the observations of one psychologist and the results of a battery of psychological tests and interviews; "blind" procedures were not used. It is obvious that such categories as restless, hot-tempered, and impulsive (none ever defined) are highly subjective and in addition are context dependent. That is, most people exhibit such behavior in some situations. Yet, the bulk of the evidence associating XYY and aggression results from the composite of many such studies, each relying on one of a few observers, a few XYYs, and a group of ill-defined categories.

Some investigators have attempted to confer more validity to their data by incorporating the double-blind techniques mentioned above into their protocols in order to avoid observer and subject bias. Noel et al.(38) claimed to be able to distinguish XYYs from non-XYY controls solely on the basis of psychological tests and interviews. Yet, the descriptions of XYYs in this study remain subjective and include terms such as "impulsiveness," "lack of emotional control," and "fits of temper." No traits are cited as distinctive for all XYYs studied, and the conclusions of the investigators simply extrapolate from an *impression:* "The most significant impression gained of the XYY subjects was their apparent inability to integrate aggression normally into their perception of reality. It would seem that, for these subjects, the aggressive drive has to be strictly controlled and can only be freely expressed in fantasy"(38; p. 392). In this way, Noel and coworkers postulated a psychological mechanism for aggression in XYY males, without ever establishing the existence of the so-called "aggressive drive."

The inadequacy of this conclusion illustrates not only the difficulties researchers have in defining aggression but the unfortunate tendency to quickly generalize the "results" of isolated studies, an error endemic in most work on XYY. Noel and coworkers report that XYYs demonstrated aggressive behavior in specific contexts: "It seems that all examined XYY subjects occasionally became aggressive, with fits of temper, and behaved impulsively when faced with frustration." Yet, the authors go on to conclude that XYYs have a personality trait, aggressiveness. Two steps are missing here. First, the context-dependent behavior is generalized. A description, such as "he behaves aggressively in some situation," is extended to a personality type, "he is aggressive." This simplistic and unsupported jump is partially concealed by defining aggression as a "trait" of individuals rather than as a description of a behavior in certain situations. Unfortunately, it is characteristic of the XYY literature that this flawed study has been cited by recent authors such as Hamerton as "proving" the existence of an XYY syndrome(45).

The third type of association between XYY and aggression relies upon an apparently more objective criterion: imprisonment for criminal actions. Since the large majority of XYYs lead normal lives, this approach is limited to those already incarcerated. Still, imprisonment as a criterion would seem to avoid the problems of definition pointed out above. Yet, even such a straightforward category is difficult to use. As the National Commission on the Causes and Prevention of Violence pointed out, of 9 million serious crimes committed during one year, less than 1 percent result in the eventual imprisonment of a criminal(46). The vast majority of crimes are either never reported or the perpetrators are never convicted or even arrested. Borgaonkar and Shah further note that "incarcerated offenders may not be representative of the much larger group officially designated as criminals. Moreover, one can hardly expect to generalize findings from confined offenders to all those who in fact engage in criminal acts"(17; p. 193). A pertinent example from the XYY literature is the study conducted by Franzke, cited by Borgaonkar and Shah(47; 17). Fifteen young XYYs were identified among boys with various problems, and, of these, 13 had conflicts with the law, 8 of whom received sentences to correctional institutions. But of the 5 upper-middle-class boys involved, only 1 was institutionalized, while 5 of 6 lower-middle-class and both lower-class boys served sentences. These data, when viewed with a wealth of other criminological studies, suggest that XYY males are subject to the same social class biases that XY males are when their incarceration is decided upon. Thus, which XYY male is found in prison probably depends more on cultural and social factors than on aggression alone.

An equally important difficulty involves not only imprisonment but also a definition of crime itself. An illustration is provided by Witkin et al.(37), among the most methodologically sophisticated of all XYY studies. They found that, for tall males born in Copenhagen in 1944-1947 who also agreed to have their chromosomes and their criminal records examined, the extra Y chromosome is associated with increased imprisonment. However, XYY males were no more likely to be incarcerated for violent crimes than XY males; in the small sample studied, most crimes were against property. One XYY male was even imprisoned for falsely reporting a traffic accident while he was intoxicated. What we mean by crime or antisocial behavior is by no means constant; rather, our definitions change from culture to culture and within a culture as time passes. Obvious examples involve changes in laws; more subtle examples are neglected statutes such as "blue laws" and laws requiring individual judgments, for example, those regulating obscenity. Further, the forces that interact to make some people criminals and others law-abiding citizens also change with time and place. Social class and ethnic components in American society are unique and constantly changing. We cannot assume that the factors discussed here interact in the same way, or at all, in different time periods or in different societies.

The realization that crime is a cultural and historical phenomenon has crucial implications for all criminologic research, especially that involving XYY, since much of this research has been done in various countries on subjects of various ages. For example, the Danish study mentioned in the preceding paragraph is really only pertinent to the population of Copenhagen during a time period characterized by a particular definition of crime and manner of dealing with suspected law breakers, as well as a unique set of social and cultural influences acting upon potential criminals. To generalize these results to other societies at other times requires that all important variables are the same in the two environments. While this requirement, if taken to an extreme, would vitiate the generalization of most psychological and sociological studies, its importance must not be overlooked because of the difficulties that arise. Especially in research on subjects, such as the XYY karyotype, in which it is not yet even known which are the important variables to worry about, one is not able to make sweeping conclusions on the basis of isolated research. The theoretical basis for this caveat comes from the nature of gene-environment interactions particularly related to the causation of behavior.

THE NATURE OF CAUSALITY

Is it possible, then, to design a study that could establish the extra Y chromosome as "the cause" of aggression? To extract the essence of this question, can we distinguish genetic from environmental factors in their influence on human behavior? To be sure, both genetic and environmental factors play a role; however, they cannot be separated by any currently acceptable method. For example, the large prospective studies in which newborns were screened and then followed cannot possibly control for different parental attitudes, school performance, and perhaps such factors as size and muscular coordination of the subjects. We can conceive of studies designed to separate genetic and environmental factors in which newborns would be reared in identical artificially controlled environments, but such a design would be logically dubious, since the social components of a behavior like aggression would be specific to the condition chosen and not necessarily applicable to the conditions in a society at a specific period. More important, moral imperatives prevent such manipulations of humans. It must be realized that the interaction of gene and environment may well be unique to a given situation. Thus, different interactions may yield different results(48). Nevertheless, a "science" termed "behavioral genetics" exists and purports to establish the genetic bases of human behavior. Given the impossibility of accomplishing this task with humans, behavioral geneticists can produce only incomplete or biased results. Studies of the XYY karyotype provide good examples of the problems that arise when genetic factors are emphasized to the exclusion of environmental concerns. To simplify the complex web of causality renders the conclusions simplistic and untrustworthy.

In addition to the claims that XYY causes a behavioral abnormality, there have been attempts to include the XYY character in the category of disease, with use of terms such as "syndrome" and even the hormone "treatments" of XYY males we referred to earlier. Again, as with behavior, our ideas of causation strongly depend upon the definition we choose for disease. To begin with the more straightforward example, if diseases are thought of and classified mainly in biological terms (such as organ pathology, biochemical pathways, and laboratory abnormalities), then emphasis will be placed on finding biological causes. This is generally the case, since the views of physicians and medical researchers are deeply rooted in both the cell theory and the germ theory, in which disease is viewed as dependent upon a small set of determinants (often only one). On the other hand, a unified or holistic definition of disease integrates multiple determinants such as social relationships, psychological development, cultural adaptation, economic status, and so forth with the organic(49, 50, and 51).

Taking this holistic approach, disease and behavior form a causal web, a complex of interrelated factors, some of which are necessary but none of which is independently sufficient for the development of a given disease. Although all of these innumerable determinants in their most reduced form are either genetic or environmental in origin, their overall effect is far from the simple sum of their individual effects. In fact, to even think in terms of individual effects is incorrect, since genes cannot determine health or disease outside of an environmental context, and environmental influences must be considered relative to the specific biological system with which they interact. This principle must be kept foremost when the study of an abnormality of the genotype such as XYY is undertaken. Most medical researchers, keen on describing and classifying yet another disease, have consistently attempted to collect the characteristics of all of the cases of the XYY karyotype that have been described in the literature under a single headline, such as "the XYY syndrome." The futility and worthlessness of these attempts should be obvious within a holistic framework of disease. Moreoever, the fact that the vast majority of XYY males are "normal" by medical, psychological, and sociological criteria argues strongly against terming this unusual karyotype a disease or even a syndrome.

THE SOCIAL SETTING OF XYY RESEARCH

As in the case of any other scientific activity, XYY research did not occur in a social vacuum. Scientists, as members of their society, train and pursue research amidst a variety of social influences, and their theories bear an indelible social imprint. As such, scientific theories cannot be said to be ethically or politically neutral; they suggest, directly or indirectly, a view of the world with specific social consequences. XYY research and the social climate of the late 1960s illustrate clearly the interplay between science and its social context.

Three interconnected trends that influenced the course of XYY research can be identified: concern with the problems of crime and violence, expanded medical efforts to alter behavior, and efforts to determine the genetic basis of human behavior. Concern with crime and violence was engendered in large part by the "crime wave" of the late 1960s and the ghetto uprisings in various cities; both seemed to forebode increased lawlessness and rejection of the social order. The extent of public concern is perhaps best illustrated by the creation of three presidential commissions to investigate such problems between 1967 and 1969.[3] In addition, increased funding was made available for research and practical aspects of law enforcement, especially through the Law Enforcement Assistance Administration (LEAA) of the Justice Department. This source of funding constitutes the most obvious link between XYY and social currents, since much XYY research was funded through the LEAA and the Center for the Study of Crime and Delinquency of the National Institute of Mental Health.[4] Predictably, XYY research was channeled to some extent toward those aspects of the abnormality that might yield information concerning the causes of violence. A few researchers explicitly recognized this influence; Jarvik and coworkers noted that "current, and well-founded, public concern over violence . . . has stimulated the search for a means not only of understanding aggression, but of controlling it as well." The authors go on to argue that "applied research aimed at curbing unnecessary and excessive violence" should follow a biological approach based on the discovery of the XYY abnormality(42; p. 675).

The second trend, the expansion of medical efforts to alter behavior, had an important if less direct influence on XYY research. The late 1960s saw an increased reliance upon medicine in dealing with social problems (such as alcoholism), individual problems (such as stress), and educational problems (such as school discipline, through the syndrome of hyperactivity) (52 and 53). Even the problem of violence was thought by some to be susceptible to a medical solution through psychosurgery (54 and 55). Within the widening medical purview, XYY research seemed to many a natural approach to the problems of aggression and crime. If "the cause" of some aggression was indeed a genetic abnormality, then aggression was a disease and should be treated as such; medical and scientific, as opposed to social solutions, should be sought.

The third trend, efforts to determine the genetic basis of behavior, also helped establish a congenial milieu for XYY theories. The science designated as "behavioral genetics" flourished in the late 1960s and centered mainly on the issue of heritability of intelligence. The IQ controversy, which attracted public attention beginning in 1969, helped to establish scientific and social respectability for behavioral genetics and to illustrate the application of this "science" to behavioral questions, for example, learning(56). The attribution of aggression to an extra chromosome thus seemed consistent with other scientific and social knowledge.

[3]The reports of these commissions were entitled *The Challenge of Crime in a Free Society* (1967), *Report of the National Advisory Commission on Civil Disorder* (1968), and *To Establish Justice, To Insure Domestic Tranquility* (1969).

[4]For example, see J. Hunt(5) on a study conducted by L. Razavi at Massachusetts General Hospital; research proposal of R. T. Rubin in J. West, Proposal for Center for the Reduction of Life Threatening Behavior of UCLA; and Summary of Active Research Grants of the Center for Studies of Crime and Delinquency, National Institute of Mental Health, July 1, 1974.

What emerges, then, is a mutually reinforcing web of theories and trends. XYY research derived support from the concern with crime and social unrest, reliance on medicine, and popularity of behavioral genetics. Concomitantly, XYY theories seemed to indicate a solution to some crime problems consistent with other scientific claims and amenable to medical control. The social consequences of XYY theories become obvious when viewed in this light: if crime and aggression are indeed biological, then social factors such as poverty and oppression that are often associated with crime can be ignored in favor of a medical solution. Disturbing issues of social injustice remain submerged, and the social order is not brought into question. A social issue becomes disguised as a scientific task.

XYY, SCIENTISTS, AND THE PUBLIC

XYY research represents a sorry episode in the history of science. It is an area of study consistently flawed by the most elementary of methodological errors, by the biases of researchers, and by a misunderstanding of genetic concepts. While all science contains considerably more of the subjective than the public is led to believe, this particular field is marred to an unusual extent. We believe that this situation may have been inevitable, when we take into account several factors that played a role in the development of XYY research. These are, first, the direction set by Jacobs' original approach and findings; second, the societal atmosphere of the late 1960s; and, third, the orientation of the researchers who chose to enter this area because of their interest in it as an approach to social problems.

This history has resulted in a situation in which it is now much more difficult, if not impossible, to do studies in an ethical way even when dealing with less controversial aspects of XYY males. Further, continued research into the connection between XYY and crime or other "deviant" behaviors, even though some earlier misconceptions may have been corrected, tends to lend credence to the concept of genetically based antisocial behavior as an explanation for many social problems. We suspect that the publicity concerning XYY research has played some role in generating the following claim, which was read by approximately 60 million Americans in their daily newspaper. It is the response of famed columnist Ann Landers to a plea for help from a discouraged parent whose son had been "on drugs" and was committed to reform school(57): "We now know that the genetic factor can be a dominating influence in behavioral patterns. Some children inherit fragile nervous systems. They go haywire and crack up—don't respond to parental love or professional help. So stop feeling guilty. You've done your best. Angels can do no more."

Clearly, XYY research and research in general into the origins of crime and aggression are, of their very nature, issues of great social import very much in the public arena. The XYY myth was promulgated widely among the public with the social consequences we have described. Yet, those scientists who recognized the defects in this research have either failed to speak up at all or have restricted their critiques to academic journals(58-60 and 17, for example). There these critiques were destined to remain, gathering dust in libraries and never reaching the outside world where the original claims still held currency. This is not surprising. Scientists have always been loathe to air their criticisms of other scientists in public, even when the issues involved were influencing social policy. For instance, there was a host of critics of eugenics theories among geneticists in the early twentieth century, but their attacks remained confined to scientific journals, while sterilization laws, miscegenation laws, and the Immigration Restriction Act of 1924 were being passed(61-64).

It was with this history in mind that we, as members of Science for the People, began, in 1974, to protest the newborn XYY screening program being carried out by

Harvard Medical School researchers at a Boston maternity hospital. This study showed the defects and problems of many of the prospective studies, including the effect of parental knowledge of the XYY variation on the children's upbringing. We brought a complaint before a Harvard committee to point out the lack of real informed consent, the possibility of stigmatization of the children, and the ways in which the design of the study rendered any conclusions about XYY behavioral patterns meaningless. We felt that the risks of this study far outweighed the very questionable benefits. Not only were we concerned about the harm being done to the families and children involved in the project, but we also saw the possibility of finally bringing to the public's attention the history of XYY research(65-68, 41, and 19).

The response of much of the academic community was to line up in defense of the study. While many scientists admitted the serious defects in the program, they were more concerned about the precedent of stopping a scientific study. As a result, one Harvard committee was divided in its opinion and another rejected our complaints. However, in the process of this struggle, we had contact with the media and held a press conference. The publicity brought other public interest groups into the controversy, and the pressure finally caused the researchers to stop their screening. The airing of the XYY issue also resulted in the halting of all other newborn XYY screening programs of which we are aware. Finally, the arguments over the study in Boston led to a number of newspaper articles over the past few years that have helped to reverse the XYY myth(69-74).

However, despite the negligible accomplishments of XYY research and the harm that it has done, studies in this area continue to be funded. We may well see in the future an attempt to reinvigorate the XYY myth. As we have shown, ideas on the biological origins of criminality have tended to resurface from time to time. Scientists who are aware of these issues should be alert to such efforts and should take the responsibility of working with the public to expose both the social assumptions and the social function of such research.

References

1. Pickens, D. K. 1968. *Eugenics and the progressives.* Nashville, Tenn.: Vanderbilt University Press.
2. Haller, M. H. 1963. *Eugenics: Hereditarian attitudes in American thought.* New Brunswick, N.J.: Rutgers University Press.
3. Kamin, L. 1974. *The science and politics of I.Q.* New York: Halsted Press.
4. Karier, C. 1976. Testing for order and control in the corporate liberal state. In *The I.Q. controversy,* eds. N. J. Block and G. Dworkin. New York: Pantheon.
5. Hunt, J. 1973. Rapists have big ears. *The Real Paper,* July 4, 1973, p. 8.
6. Lombroso, C. 1897. The heredity of acquired characteristics. *Forum* 24:200.
7. Lombroso, C. 1911. *Criminal man.* New York: G. P. Putnam's and Sons.
8. Ellis, H. 1897. *The criminal.* 2nd ed. London: Later Scott Ltd.
9. Hooton, E. 1939. *The American criminal: An anthropological study.* Cambridge, Mass.: Harvard University Press.
10. Terman, L. M. 1916. *The measurement of intelligence.* Boston: Houghton Mifflin Company.
11. Laughlin, H. H. 1922. *Eugenical sterilization in the United States.* Chicago: Psychopathic Laboratory of the Municipal Court of Chicago.
12. Estrabrook, A. H. 1916. *The Jukes in 1915.* Washington, D.C.: Carnegie Institute of Washington.
13. Goddard, H. H. 1912. *The Kallikak family: A study on the heredity of feeblemindedness.* New York: Macmillan.
14. Sutter, J. 1950. *L'Eugénique.* Paris: Presses Universitaire de France.
15. Mark, V., Sweet, W., and Ervin, F. 1967. Letter to the editor. *Journal of the American Medical Association* 201:895.
16. Eysenck, H. J. 1964. *Crime and personality.* Boston: Houghton Mifflin Company.
17. Borgaonkar, D., and Shah, S. 1974. The XYY chromosome, male—Or syndrome. *Progress in Medical Genetics* 10:135-222.

18. Hook, E. B. 1973. Behavioral implication of the human XYY genotype. *Science* 179:139-50.
19. Beckwith, J., and King, J. 1974. The XYY syndrome: A dangerous myth. *New Scientist* 64:474-76.
20. Sandberg, A. A., Koepf, G. F., Isihara, T., and Hauschka, T. S. 1961. An XYY human male. *Lancet* 2:488-89.
21. Jacobs, P. A., Brunton, M., Melville, M. M., Brittain, R. P., and McClemont, W. F. 1965. Aggressive behavior, mental subnormality, and the XYY male. *Nature* 208:1351-52.
22. Price, W. H., and Whatmore, P. B. 1967. Criminal behavior and the XYY male. *Nature* 213:815.
23. Neilsen, J., and Tsuboi, T. 1969. Intelligence, EEG, personality deviation and criminality in patients with the XYY syndrome. *British Journal of Psychiatry* 115:965.
24. *Newsweek.* 1970. Congenital criminals. *Newsweek* 75:98-99.
25. Otto, J. H., and Towle, A. 1973. *Modern biology.* New York: Holt, Rinehart and Winston.
26. Engel, E. 1972. The making of an XYY. *American Journal of Mental Deficiency Research* 77:123-27.
27. Freedman, A. M., Kaplan, H. I., and Sadock, W. I. 1972. *Modern synopsis of comprehensive textbook of psychiatry.* Baltimore: William and Wilkins.
28. Royce, K. 1973. *The XYY man.* New York: Avon.
29. Katz, J. 1972. *Experimentation with human beings.* New York: Russell Sage Foundation.
30. Money, J. 1970. Use of an androgen depleting hormone in the treatment of male sex offenders. *Journal of Sex Research* 6:165-72.
31. Blumer, D., and Migeon, C. 1975. Hormone and hormonal agents in the treatment of aggression. *Journal of Nervous and Mental Disease* 160:127-37.
32. Glass, H. B. 1971. Science: Endless horizons or golden age. *Science* 171:23-29.
33. Montagu, A. 1968. Chromosomes and crime. *Psychology Today* 1(9):43-49.
34. Montagu, A. 1976. *The nature of human aggression.* New York: Oxford University Press.
35. Goodman, R. M., Miller, F., and North C. 1968. Chromosomes of tall men. *Lancet* 1:1318.
36. Walzer, S. 1974. Research grant proposal to the National Institute of Mental Health. A document obtained under the Freedom of Information Act by the Children's Defense Fund.
37. Witkin, H. A., Mednick, S. A., Schulsinger, F., Bakkestrom, E., Christiansen, K. O., Goodenough, D. R., Rubin, K., and Stocking, M. 1976. Criminality in XYY and XXY men. *Science* 193:547-55.
38. Noel, B., Duport, J. P., Revil, D., Dussuyer, I., and Quack, B. 1974. The XYY syndrome: Reality or myth? *Clinical Genetics* 5:387-94.
39. Wilson, E. B. 1952. *An introduction to scientific research.* New York: McGraw-Hill.
40. Zeuthen, E., Hansen, M., Christiansen, A. L., and Nielsen, J. 1975. A psychiatric-psychological study of XYY males found in a general population. *Acta Psychiatrica Scandinavica* 51:3-18.
41. Beckwith, J., Elseviers, D., Gorini, L., Madansky, C., Csonka, L., and King, J. 1975. Harvard XYY study. *Science* 187:298.
42. Jarvik, L. F., Klodin, V., and Matsuyama, S. S. 1973. Human aggression and the extra Y chromosome: Fact or fantasy? *American Psychologist* 28(8):674-82.
43. *Lancet.* 1974. Editorial: What becomes of the XYY male? *Lancet* 2:1297-98.
44. Owen, D. R. 1972. The 47, XYY male: A review. *Psychological Bulletin* 78:209-33.
45. Hamerton, J. L. 1976. Human population cytogenetics: Dilemmas and problems. *American Journal of Human Genetics* 28:107-22.
46. National Commission on the Causes and Prevention of Violence. 1969. *To establish justice, to insure domestic tranquility.* Final report of the Commission. Washington, D.C.: U.S. Government Printing Office.
47. Franzke, A. W. 1972. Sociology and cytogenetics: A study of the stigmatization or non-stigmatization of fifteen XYY karyotypic individuals in differentiated subcultural typologies. An unpublished masters thesis, University of Missouri at Kansas City.
48. Feldman, M. W., and Lewontin, R. C. 1975. The heritability hang-up. *Science* 190:1163-68.
49. Engel, G. C. 1960. A unified concept of health and disease. *Perspectives in Biology and Medicine* 3:459-85.
50. Fabrega, H., Jr. 1974. *Disease and social behavior: An interdisciplinary perspective.* Cambridge, Mass.: MIT Press.
51. Parson, T. 1958. Definitions of health and illness in the light of American values and social structure. In *Patients, physicians and illness,* ed. E. G. Jaco. Glencoe, Ill.: Free Press.
52. Waitzkin, H., and Waterman, B. 1974. *The exploitation of illness in capitalist society.* New York: Bobbs-Merrill.
53. Schrag, P., and Divoky, D. 1975. *The myth of the hyperactive child.* New York: Dell.
54. Chorover, S. L. 1973. Big Brother and psychotechnology. *Psychology Today* 7(5):43-54.

55. Chorover, S. L. 1974. Big Brother and psychotechnology II; The pacification of the brain. *Psychology Today* 7(12):59-69.
56. Block, N. J., and Dworkin, G., eds. 1976. *The I.Q. controversy*. New York: Pantheon.
57. Landers, A. 1975. A father's lament. *The Boston Globe,* November 21, 1975, p. 28.
58. Coleman, L. S. 1974. Perspectives on the medical research of violence. *American Journal of Orthopsychiatry* 44:675-87.
59. Goldstein, M. 1974. Brain research and violent behavior. *Archives of Neurology* 30:1-23.
60. Fox, R. G. 1971. The XYY offender: A modern myth? *Journal of Criminal Law, Criminology and Police Science* 62:59-73.
61. Ludmerer, K. H. 1972. *Genetics and American society*. Baltimore: Johns Hopkins University Press.
62. Allen, G. 1974. A history of eugenics in the class struggle. *Science for the People* 6(2):32-39.
63. Provine, W. B. 1973. Geneticists and the biology of race crossing. *Science* 182:790-96.
64. Beckwith, J. 1976. Social and political uses of genetics in the United States: Past and present. *Annals of the New York Academy of Sciences* 265:46-58.
65. Elseviers, D. 1974. XYY: Fact or fiction? *Science for the People* 6(5):22-24.
66. The Genetic Engineering Group. 1975. The XYY controversy (continued). *Science for the People* 7(4):28-32.
67. Culliton, B. J. 1974. Patients' rights: Harvard is site of battle over X and Y chromosomes. *Science* 186:715-17.
68. Culliton, B. J. 1975. XYY: Harvard researcher under fire stops newborn screening. *Science* 188:1284-85.
69. Brody, J. E. 1974. Scientists' group terms Boston study of children with extra sex chromosome unethical and harmful. *The New York Times,* November 15, 1974, p. 16.
70. Greenberg, J. 1975. Genetic witch hunt: Blaming it all on biology. *Tallahassee Democrat,* February 3, 1975.
71. Knox, R. A. 1975. XYY battle is over, research ethic war isn't. *The Boston Sunday Globe,* July 6, 1975, p. A4.
72. Beckwith, J. 1976. XYY syndrome. Letter in *Newsweek* 87(7):4-5.
73. Okie, S. 1976. XYY: An ethical issue in a genetic riddle. *The Washington Post,* July 25, 1976, pp. C1, C4.
74. Brody, J. E. 1976. A chromosome link to crime is doubted. *The New York Times,* September 12, 1976, p. 18.

Reed Pyeritz is a resident in medicine at the Peter Bent Brigham Hospital in Boston. *Herb Schreier* is a child psychiatrist at the Fernald School and the East Boston Health Clinic. *Chuck Madansky* is a graduate student in virology at the University of Massachusetts Medical School, Worcester. *Larry Miller* is a third-year medical student at Harvard Medical School, Cambridge, Mass. And *Jon Beckwith* teaches and does research in the department of microbiology and molecular genetics at Harvard Medical School.

Political Determinants of Violence

Richard Kunnes

Many authors look for some of the causes of individual aggression in the context of a debate over social versus biological determinants of aggression. They often conclude that there are only social causes of aggression, albeit in an individualistic framework.

My use of the term *aggression* in this paper is as a synonym for *violence,* when violence is defined narrowly as an action or a form of behavior that results in physical damage to one or more humans. I wish to examine the terms of the question, what is, if any, the importance of the biological determinants of aggression? or even, are there biological determinants of aggression? I think two words within these questions, namely the words *biological* and *aggression,* ask the wrong question and/or tend to limit the scope of the answer. In effect, by narrowly defining the scope of the debate to the genetics and/or psychology of isolated individuals and their acts of aggression, by only dealing with an individual's set of genes and/or personality, the question generates a form of reactionary or repressive political ideology by avoiding or ignoring the political context in which most violence occurs. It is a repressive ideology because the question forces us to focus on the blander aspects of aggression, such as psychological aggression, as opposed to something as blatantly violent as genocide. The question tends to push us to examine what goes on in the psychodynamics of isolated individuals' heads, as opposed to the political dynamics in international corporate headquarters. To push the genetic question, then, from my point of view, is to be both ahistorical and apolitical, two key aspects of a repressive ideology.

If I appear less than excited by biological or psychological explanations of history—in general—and violence and aggression—in particular—ask yourself what a more appropriate, informative, and constructive way to view violence and aggression is.

When liberal academicians talk of the more apparent and extreme forms of aggression or violence, they usually only examine individual acts, such as one person killing another person. But individual acts of murder, as tragic as they are, make up only a very small aspect of fatal violence in the United States and in the world, as compared to institutional[1] acts of violence. For example, there were fewer than 25,000 individual acts of murder last year in the United States(1; pp. 43-87; 2). There were, however, 50,000 fatalities from auto accidents(3; pp. 52-57), 250,000 fatalities due to alcoholism(4; vol. I, pp. 87-93), and an estimated 480,000 cancer and cardiopulmonary fatalities caused by air, water, and food contaminants and pollutants(4; vol. II, pp. 101-5). The United States Public Health Service estimates that over half a million citizens die each year solely because of the inaccessibility or the nonexistence of conventional medical services(4; vol. II, pp. 182-84).

It seems more than coincidental that in studies of violence these other causes of violent fatalities, such as auto accidents, alcoholism, pollution/contamination, and inadequate health care are not even viewed as forms of violence. They are simply defined out of the picture. They are not allowed to be an issue in any debate about violence, because, if they were, the culprits and criminals would not be an individual's genes and/or psychopathology but a class of corporate elites who control our society. Some of the overriding causes of auto accidents are, for example, hazardous crowding from too many cars, dangerously designed cars, and insufficient or nonexistent mass-transit systems. Who's responsible? It seems as though it is some form of an auto-oil-highway-industrial complex whose economic interests in a political context, rather than any genetic makeup in a psychobiological context, insure an ever-mounting source of violence.

When considering alcoholism, we must remember from where alcohol comes. It comes from major corporate distilleries and distributors, which, in turn, are subsidiaries of major transnational conglomerates with important links to the Defense Department. Even the American Medical Association (AMA) is a major shareholder in two major breweries in the United States, namely Anheuser-Busch and Pabst(5; pp. 130-35). The AMA has a clear conflict of interest concerning alcohol addiction. One, they own stock in alcohol, and, two, since alcoholism is the most commonly recognized major source of morbidity in the United States, alcoholism creates an unending supply of patients for physicians.[2]

Consider also air, water, and food pollutants and contaminants. The United Nations World Health Organization says these contaminants are the cause of 85 percent of *all* cancer(6; pp. 35-102). These are corporately produced contaminants and thus corporately produced cancers, corporately produced violence. The world's major transnational conglomerates are the greatest single cause of cancer in the world.

The pollution and contamination picture appears to be worsening, because, as industrial concentration and monopolization increase, so does cancer. Larger industries and factories produce far more waste by-products. Industrial food processing requires longer storage periods and shelf times for higher profits, and so foods contain more carcinogenic preservatives and contaminants. Monopolization and concentration of industrial processes have led to monopolization and concentration of the control of political processes. For example, the federal regulatory agencies have increasingly become the protectors of the major conglomerates rather than of the consumer(7; pp.

[1]The word *institutional*, as used in this paper, describes a complex series of broad social events and relationships, including both cultural and corporate forces. While *institutional* versus *individual* is a somewhat false dichotomy, for purposes of clarity I hope it is useful. For example, if someone strangles her or his next-door neighbor, I am calling this an "individual act," with no *overt* institutional relationships; that is, the murder can occur without any overt intermediaries. On the other hand, auto fatalities require social intermediaries such as complex road systems, complex manufacturing processes, the culture of cars, commercialism, and so on, all of which fundamentally involve and have powerful and overt institutional roots.

[2]While not that many physicians earn any significant income from the treatment or cure of alcoholism per se, though there are an ever-increasing number of private alcoholic hospitals, physicians do earn plenty treating the results of alcoholism, for example, ulcers, liver disease, heart disease, neurological diseases, and so forth.

82-201). Thus, the incidence of cancer has been rising year after year(8; pp. 48-93). And yet death by cancer is not, at least academically, studied as an example of aggression or violence. No one would suggest, as they frequently do with individual violence (e.g., XYY research) that studies be made of the genetic makeup of the members of the boards of directors of these corporations. The issue, however, is not corporate leaders as individuals but rather their political system's ability to prevent us from challenging and changing their institutional, and, most importantly, their class power and privilege.

Let us now look at the issue of nonexistent or inaccessible health care. In spite of the fact that the United States spends far more than any other country on health care (10 percent of the gross national product [9; pp. 127-38]), most of us receive second-rate health care, and some of us receive none. Life expectancy in this country is below at least 12 and as many as 17 other countries(10; pp. 40-45). If we spend so much on health care, why do we have such poor health? Let me mention just two small but typical examples. The financial resources for medical care are increasingly shunted away to the heavy manufacturing priorities of a medical-industrial complex. There's far more profit in esoteric, expensive, complex equipment than in conventional, accessible, prevention-oriented out-patient services. As an example, there are almost 800 hospitals in the United States equipped and staffed to perform open-heart surgery, but at least one-third of them have never performed a single open-heart procedure(11; pp. 33-49). There is more profit for the major conglomerates in the *purchase* of equipment than in the provision of services. There are only limited numbers of trained health personnel, in part because they are misutilized and disproportionately utilized in the staffing of such narrowly relevant services, services that have only limited applicability. This malapportionment of money, staff, and equipment results in a lot of deaths, but such are the priorities of the system.

Even the medical care that *is* provided is often fatally inappropriate, however profitable. Last year, it was estimated that there were in the United States, as many as 3 million *unnecessary* surgical operations that resulted in at least 25,000 unnecessary deaths(12; pp. 5-185). You cannot get much more overtly aggressive or violent than that. But try to get the AMA or the medical-industrial complex to make itself publicly accountable and responsible or, most important, to support socialized health care in which people's lives, rather than corporate profit, are the first priority.

Who or what has the overriding responsibility for death from automobiles, alcoholism, pollution, and contaminants, and inaccessible or inappropriate health care? It is the major transnational conglomerates that put profits before lives. Economic priorities inevitably determine violence; the biological makeup of the victims does not. Blaming the victim has long been an ideological ploy of the ruling class to shift the blame away from itself.

Finally, there is violence on a broader scale: war. The whole history of modern war is not a psychobiological phenomenon; it is rooted in various forms of class conflict. The United States government performs the major managerial/bureaucratic functions for the corporate ruling class. Its essential duties are, first, to insure domestic tranquility through the maintenance of cultural and ideological hegemony. Universities, television, and movies also play a significant role here. This tranquility is necessary to have relative peace at home while economic and military wars are pursued abroad. The second duty is to constantly acquire and control new markets and raw materials by any and every means available, including the ultimate form of violence—genocide. The history of the United States is primarily a history of violent colonial and neocolonial expansion and control, in a word, imperialism. The institutional violence of American imperialism has been with us since America's so-called discovery by white

Europeans, from its destruction of native American nations to the government's use of genocide in Indochina.

The scope, intensity, and variety of the violence appear to be getting at least potentially a bit worse. The 1976 Department of Defense budget is over 100 billion dollars. While weapons are often used for our colonial expansion abroad and domestic control at home, the actual federal expenditures for them also serve a somewhat different purpose, the stimulation of the economy. To the capitalist ruling class, such military spending has a number of advantages over more socially relevant, though less violence prone, kinds of spending.

For military contractors, weapons spending means profitable sales without having to compete for consumer dollars. Weapons spending is more appealing to employers than expenditures for higher welfare or unemployment benefits—both of which might reduce workers' incentives to work for increasingly inflation-gutted wages.

No other government program contains all these so-called advantages. No other program can count on such reliable support. The support from hard-core promilitary conservatives, combined with so-called "pragmatic" moderates who want to provide jobs and stimulate the economy—no matter how it is done and at whose expense—add up to an effective congressional majority for the Pentagon's budget.

Ever-increasing weapons expenditures appeal to liberals as well. The 1958-1960 recession, which until recently was the most serious economic slump since the 1930s, was ended through President John Kennedy's increases in military spending. Kennedy, the most liberal president since the Second World War, could only stimulate the economy by inventing a "missile gap" that had to be closed and by intensifying the arming for Vietnam.

Recently, President Gerald Ford has repeated the Kennedy strategy for combating recession and depression with weapons expenditures. Powerful as our present economic and political system may be, it cannot seem to produce peace *and* jobs at the same time. Such a strategy of weapons proliferation just incidentally increases the danger of blowing us all up—violently. *Every* president from Washington to Ford has presided over an expanding military budget that supports an increasingly violent power that is used to conquer and to kill our less well armed and less greedy neighbors.

Does it really make any sense to ask whether United States presidents have been genetically and/or psychologically violence prone, in the context of the historical, political, and economic direction of United States global expansion? Does anyone seriously think that, if we had a genetically different president, the general thrust of American imperialism would have been permanently thwarted and less violent? Do the genes of corporate leaders predispose them to violence, or was it more likely their personal exposure to ruling-class privilege and ideology, or are they unwitting or witting participants in a system of institutional violence?

If one wanted to engage in a kind of science fiction, one could somehow imagine that the genes of the ruling class *are* violence prone. After all, much of their wealth and power *is* passed from one familial generation to the next, with considerable inbreeding (much more than in the working class). But, if we did accept the notion that ruling-class genes are violence prone (as some people have argued in reference to acts of violence by members of another class), then perhaps it would be futile to argue with such people; they should simply be arrested and imprisoned as a menace to society.

If institutional violence were genetically determined, it seems as though it would be more randomized and not as predictable and systematic as it is. For example, United States intervention and violence in Indochina, Iran, Cuba, Korea, Chile, Portugal, Angola, and so on, was totally predictable and completely consistent with our national imperialist policy. That policy can be fully understood without resorting to biological or psychological explanations.

Since the Second World War, there has not been a *single internalized struggle* between socialist-oriented and capitalist-oriented forces in which the United States did not *violently intervene* on behalf of the capitalist-oriented forces. None of these countries constituted a threat to the safety of the United States, only to its control of foreign markets and resources.

What about other sources of death that are not recognized as violence? Millions of people die unnecessarily from malaria, yellow fever, parasitic infections, and countless other preventable diseases. The single most important cause of these deaths are the conditions created by the impoverishment and emiseration caused by colonialization and neocolonialism, euphemistically called an import-export trade deficit.

Millions of people die every year from malnutrition in spite of the planet's ability to feed much more than twice its current population. The *single most important factor* causing widespread deaths due to malnutrition is, in the words of Barry Commoner, "colonial exploitation . . . the exploitation of poor nations by rich ones . . .," currently and predominantly the United States(13; p. 59). Commoner suggests that the only life-saving solution is a redistribution of wealth—not charitable handouts, but returning to those impoverished nations all we have violently taken from them.

America's corporate wealth has continued to grow only at the life-and-death expense of others. Increasingly, the most impoverished nations and peoples of the world will resist United States violence. And, if only out of a sense of survival, these people will fight violence with violence.

References

1. U.S. Federal Bureau of Investigation. 1974. *Uniform crime reports.* Vol. 46. Washington, D.C.: U.S. Government Printing Office.
2. Henry, A. F. 1974. *Suicide and homicide.* New York: Macmillan.
3. National Safety Council. 1974. *Accident facts.* Chicago: National Safety Council.
4. U.S. Division of Vital Statistics. 1974. *Vital statistics of the U.S., 1974.* Vols. I and II. Washington, D.C.: U.S. Government Printing Office.
5. Kunnes, R. 1972. *Your money or your life: The medical marketplace.* New York: Dodd, Mead and Company.
6. United Nations World Health Organization. 1974. *The work of WHO—Annual report to the director general to the World Health Assembly and to the United Nations.* Geneva: World Health Organization.
7. Turner, J. S. 1970. *The chemical feast.* New York: Grossman.
8. U.S. National Institutes of Health. 1975. *Atlas of cancer mortality for U.S. citizens: 1950-1969.* U.S. Department of Health, Education, and Welfare publication no. NIH 75-780. Washington, D.C.: U.S. Government Printing Office.
9. Fuchs, V. R. 1974. *Who shall live: Health economics and social choice.* New York: Basic Books.
10. National Center for Health Statistics. 1974. *Vital and health statistics.* Series 7, no. 2. Washington, D.C.: U.S. Department of Health, Education, and Welfare.
11. Ribicoff, A. 1973. *The American medical machine.* New York: Saturday Review Press.
12. Williams, L. P. 1972. *How to avoid unnecessary surgery.* New York: Paperback Library.
13. Commoner, B. 1975. How poverty breeds overpopulation (and not the other way around). *Ramparts* 13:21-25, 58-59.

The Environmental Crisis

The late 1960s saw the emergence of the ecology movement as a major political force. Recognition of increasing pollution, waste of resources, and a declining quality of life led people to call for a new kind of relationship to the natural world. One of the catalysts for this movement was Paul Ehrlich's *The Population Bomb*(1). Ehrlich argued that the fundamental cause of our environmental crisis is the world's increasing population. He called for massive efforts to lower birth rates, voluntarily if possible, coercively if necessary.

Garrett Hardin has extended this philosophy by suggesting that we should withhold food from countries that continue to have high birth rates so as to discourage reproduction(2). He argues that we must view the world as a lifeboat that is not large enough to hold everyone, so that some persons must be left to drown so that others will survive.

These views of environmental destruction as results of human reproductive patterns and the resulting exponential growth of populations are now widely held in our society. They have been supplemented by the computer modeling of *The Limits to Growth,* which sees the world as a large system the behavior of which can be represented mathematically (3). *The Limits to Growth* model predicted worldwide catastrophe unless growth is severely restricted and suggested that planning on a global scale is necessary to save our planet.

These approaches see the basic problem as the growth and development of human populations on a finite earth and argue that the human species must ultimately be limited by the same forces that control animal species. Just as we see animal populations competing for scarce resources in nature, so too are we destined to suffer more and more acute crises as population grows and resources become scarcer. The analogy between human and animal populations is common, and it underlies much environmentalist thinking. Thus, much of our approach to social problems, such as starvation and pollution hinges on our view of the natural world.

References

1. Ehrlich, P. 1968. *The population bomb.* New York: Ballantine Books.
2. Hardin, G. 1974. Living on a lifeboat. *BioScience* 24:561-68.
3. Meadows, D. H., Meadows, D. L., Randers, J., and Behrens, W. W., III. 1972. *The limits to growth.* New York: Universe Books.

Ecological Determinism

John Vandermeer

The environmental crisis has matured somewhat since the sixties. This maturation process seems to have changed crisis into paradox. In case after case, ecological crises can more appropriately be called ecological paradoxes.

For example, on the northern shore of Lake Superior, the Reserve Mining Company provides jobs for several thousand people, yet, at the same time, it dumps taconite tailings contaminated with asbestos-like fibers into Lake Superior, potentially poisoning the water supply of the surrounding area. According to company officials, cessation of the dumping part of the operation would make it impossible for the company to operate in the black. Good business sense would require the company to move its operation from its present site, causing a loss of several thousand jobs(1).

To cite another example, William T. Huston, president of Watson Industrial Properties, said in a speech delivered before the Southern California Corporate Planners Association, in 1974(2; p. 274):

One of the facts of life that motivates me as chairman of the Los Angeles Chamber of Commerce is that the five-county Los Angeles area needs to create 75,000 jobs each year just to take care of existing labor ranks, without any migration into the area. Failure to do so will have serious social and economic repercussions. If all current environmental regulations were to be put into effect as they stand now, approximately 800,000 jobs would be lost in this area.

This pattern is repeated time and time again. The environmentalist says that we must stop polluting; the industrialist says that if we clean up we will go bankrupt. The sad fact is that they both seem to be right.

To detail the extent of our current ecological problems would be superfluous at the present time. The press fully documents such problems, and we can watch their daily progression on the evening news. No one would deny that every major United States city suffers from serious air pollution, that many major rivers are polluted, or that a majority of the world's people do not get enough to eat. Clearly, our ecological problems are very severe.

Yet, we also see the disturbing conflicts that accompany each and every "ecological crisis." Air pollution in many major cities reaches dangerous levels, but the auto industry lobbies to defer legislation on emission controls. People starve in Latin America, but the secretary of agriculture notes that food is part of our "negotiating package"(3; p. 279). Every ecological "problem" seems to carry with it something structural that renders the problem insoluble, at least in the near future. It is certainly an ironic situation that freedom for the average woman and man seems to be freedom to choose between ecological ruin or economic ruin, death by poison or death by starvation. People will insist on an understanding of the cause of a contradiction that affects their lives so directly.

But, by the time the paradoxical nature of the environmental crisis became obvious to the public, an apparent cause of that fundamental contradiction had been firmly entrenched in the public's consciousness. The basic problem was seen as *too many people.* If there were not so many people in Latin America, Latin Americans would have enough food to go around. If there were not so many people in the United States, the staggering number of automobiles would no longer exist. Virtually every ecological and environmental problem could be, and was, plausibly tied to the spector of *too many people.*

As the "population problem" became central to the ecological movement, a host of important factors were swept under the rug. Many of these factors are beginning to reemerge, but, in the minds of many people, the so-called population problem still remains an important causal factor in the generation of the sort of ecological paradoxes described above.

From the seemingly universal acceptance of the simplistic "population bomb" arguments of the late sixties, it seems that there have emerged two somewhat more sophisticated analyses. I will call these the *reactionary analysis* and the *liberal analysis.*

REACTIONARY ANALYSIS

This line of attack probably began in modern times with the Paddocks' book *Famine—1975* (4), in which it is suggested that the triage system of medical practice on a fighting front be modified to apply to countries. Under a triage system, those wounded in battle are placed in three separate categories: those who will survive with or without medical attention, those who are likely to die even with medical attention, and those who are likely to survive only if medical attention is given. Efforts are concentrated on saving those in the last category. Similarly, the Paddocks suggest that food aid should be given only to countries that have some hope of solving the "food-people gap." Those countries deemed "too far gone" in terms of their population explosion would be denied aid under such a plan.

This concept leads naturally to the recent, more carefully thought-out position of Garrett Hardin. Hardin's analysis is in fact more complicated and sinister than many of his detractors admit. It proceeds along two interrelated lines. First, his lifeboat ethics, as he states in *BioScience*(5; p. 561):

Metaphorically, each rich nation amounts to a lifeboat full of comparatively rich people. The poor of the world are in other, much more crowded lifeboats. Continuously, so to speak, the poor fall out of their lifeboats and swim for a while in the water outside, hoping to be admitted to a rich lifeboat, or in some other way to benefit from the "goodies" on board. What should the passengers on a rich lifeboat do? This is the central problem of "the ethics of a lifeboat."

Second, his theory of the guardians of civilization(6; p. 1297):

It is unlikely that civilization and dignity can survive everywhere; but better in a few places than in none. Fortunate minorities must act as the trustees of a civilization that is threatened by uninformed good intentions.

This two-step analysis is backed by three implicit assumptions: first, that the world's population, because of its excessive size, is locked into a scarce-resource situation, and second, that the distribution of resources into have and have-not lifeboats is inevitable, and third, that privileged classes are necessary to make a better life (or at least an acceptable life) for future generations.

Hardin's analysis (the lifeboat model plus the trustees of civilization theory), coupled with his three assumptions (population causes scarcity, class structure is inevitable, class structure is desirable), leads to the conclusion that it is our moral responsibility to future generations to withhold resources from those who presently do not have enough to survive. If we do not, they will only produce more babies, thereby exacerbating an already critical situation. As the *National Observer* put it, "let 'em starve" is an ethical consequence of Hardin's lifeboat ethics, and the "'em" are those who are not the trustees of civilization(7). I have argued elsewhere that Hardin's entire model is incorrect(8).

By way of a footnote I might point out that these analyses are not at all new. For example, in 1925, a prominent ecologist wrote:

[We have] an annual increase in population of nearly nine hundred thousand souls. The difficulty of feeding this army of new citizens must grow greater from year to year and ultimately end in catastrophe, unless ways and means are found to forestall the danger of starvation and misery in time.

And, in the same work:

The productivity of the soil can only be increased within defined limits and up to a certain point. The necessities of life, however, are rising more rapidly than the numbers of the population. . . . It is therefore illusory to hope that an increase in production can provide the basis for an increase in population.

Furthermore, he writes:

Instead of expanding geographically, instead of exporting men, the white race has exported goods and so has built up a world economic system, the characteristic of which is that in Europe, and lately also in America, is to be found a gigantic world scale level of manufacture while the rest of the world provides vast markets and sources of raw materials. The white race can . . . only maintain its position so long as differences in the standard of living persist in various parts of the world. Depending on their innate capabilities the various nations have taken differing steps to safeguard this predominant position.

Finally:

Everything we admire on this earth today—science and art, technology and inventions—is only the creative product of a few peoples [who] are originally perhaps of one race. On them depends the existence of this whole culture. If they perish, the beauty of this earth will sink into the grave with them.

The work is *Mein Kampf* and the ecologist is Adolf Hitler(9; p. 61; 10; pp. 121 and 155; 9; p. 25).

That the Hardin line appears so similar to the Hitler line is worth considerable reflection. If scarcity is caused by overpopulation and the population will inevitably be divided into haves and have-nots and it is a desirable thing to have such a class structure, what must we conclude? Does it not make sense that we must restrict population increase or perhaps even lower the size of the present population? Which segment of the population should that restriction hit the hardest? Those without the capability (or desire) to preserve civilization and dignity? Who are those? What makes them that way? Indeed, one might argue that the principle difference between Hardin and Hitler is that Hitler carried the arguments to their logical conclusions. Hardin stops short.

But the moral repugnance of this line of argument need not concern us. The argument is wrong. It is wrong because all of its fundamental assumptions are wrong. Let us briefly consider those three assumptions.

To assert that "overpopulation" (or a tendency to overpopulate) is a principal cause of scarcity requires a profound ignorance of elementary economic and political facts. The argument usually follows along the lines originally developed by Thomas Malthus. Populations increase at an exponential rate; requisites of life, if they increase at all, increase arithmetically. Therefore, population will necessarily outstrip requisites for life. For example, suppose a certain population lives on an island of 100,000 acres. Suppose that the land will produce in such a way that one person's necessities can be produced by 100 acres. Through the normal course of technological development, every 30 years ten fewer acres are required to meet the necessities of life for one individual. Thus, if we begin in 1970 (say) with 100 people and a requirement of 100 acres per person, all the people can be easily accommodated (100 people X 100 required acres = 10,000 acres required). If the population doubles every 30 years, by the year 2000, there will be 200 people and the requirements per person will be 90 acres per person (100 - 10). Thus, all the people could still be easily accommodated (200 X 90 = 18,000 acres required). By the year 2030, there will be 400 people and the requirements per person will be 80 acres per person (90 - 10). Still all can be accommodated (400 X 80 = 32,000 acres required). By the year 2060, there will be 800 people and a 70-acre-per-person requirement (800 X 70 = 56,000 acres required). In the year 2090, 1,600 people and 60 acres per person mean that 96,000 acres are required—close to the brink of disaster since the island only has 100,000 acres available. Then, in 2120, the population will be 3,200 people, each requiring 50 acres, leading to a total requirement of 160,000 acres, more than one and a half times the amount actually available. With 3,200 people and an island of 100,000 acres, each person will have only 31.3 acres available, while we have stipulated that 50 acres are required. What a condition of scarcity! How rapidly it came upon us!

The above argument, while the details differ from writer to writer, forms the basis for the myriad forms of the overpopulation argument. As long as the rate of technological development (as used in the restrictive sense above) is not forever greater than the rate of population growth (which it cannot be, given a finite world, they argue), overpopulation will inevitably lead to scarcity. A mathematical certainty.

How does this rather abstract analysis correspond to the real world? For example, consider the country of Guatemala, largely an agricultural economy. (Most of the Guatemalan history here is taken from Melville and Melville [11]). Throughout the 1940s, the amount of land available to the average peasant farmer slowly decreased. Many peasant farmers had to subsist on less than an acre of land. But in the early fifties, land became more easily available to peasant farmers. Fewer and fewer were forced to exist on plots as small as they had lived on prior to 1950. Then, suddenly, in 1954, land was again extremely scarce, even more so than before 1950. A neo-Malthusian might be quick to conclude that rising population throughout the 1940s caused a relatively scarce land situation; that, then a leveling off of the population in the early fifties somewhat ameliorated the problem, while a sudden burst of population growth prior to 1954 (or perhaps the coming of age of an earlier "baby boom") again created a scarce resource situation. Indeed, if one looks at the population history of Guatemala, one finds a continual increase in population that began shortly after the

Spanish conquest. Such a superficial examination tends to confirm the claims of the population bombers, if not in detail, certainly in broad outline. Just as expected, a rising population will inevitably lead to a situation of scarce resources, and Guatemala would seem to be a prime example.

But, if we take off our political blinders and look more closely, we begin to see a slightly different story. Throughout the thirties and during the early forties, Guatemala was ruled by the dictator Ubico, whose friendliness with the United States and its concerns enabled the United Fruit Company, among others, to establish a firm position in his country. Rising expectations, stemming largely from a growing bourgeoisie, forced a revolution in 1944. The left-leaning Juan Jose Arévalo was elected to the presidency and immediately began pushing for an agrarian reform that would put much of the uncultivated lands of the large landowners into the hands of the peasants. His successor, Jacobo Arbenz Guzman, accelerated the program of agrarian reform. During the early fifties, 603,615 hectares were expropriated from large landowners and put into the hands of the peasants. (This is estimated to be only 16.3 percent of the uncultivated lands held in reserve by the large landowners.)

At that time, the United Fruit Company owned more than 500,000 acres, much of which was not in cultivation but was held in reserve. (For example ". . . on the Atlantic coast they had only 4,000 acres planted and had a reserve of 88,000 acres. Even if they doubled their production and did not effect flood-fallow, they had enough reserve land for 110 years more" [11; p. 82].) The years of 1953 and 1954 were troubled for the United Fruit Company. On March 2 and 5, 1953, 209,842 acres were taken from their holdings in Escuintla. On October 17, 1953, 5,900 acres were expropriated from their properties in Suchitepequez. On February 25, 1954, 111,159 acres were taken from their holdings in Isabal. Thus, in about a year, 326,901 acres of their 500,000-plus acres were expropriated.

During this time, John Foster Dulles was United States secretary of state. Mr. Dulles said, "The Guatemalan Government boasts that it is not a colony of the United States. We are proud that Guatemala can honestly say that. The United States is not in the business of collecting colonies. The important question is whether Guatemala is subject to communist colonialism"(11; p. 96). Now, the fact that Mr. Dulles was a senior partner of the Cromwell-Sullivan law firm, United Fruit Company's lawyers, probably had nothing to do with his position regarding this issue. He probably did not even know that they had just lost 326,901 acres of their reserve land. Like all secretaries of state, he was only interested in fighting communism, which is why in 1953 and 1954 the United States openly supplied arms to the governments of Honduras and Nicaragua and to Guatemalan exiles in several Central American countries. Further, in 1954, the United States supplied arms and first-hand training to the forces of Castillo Armas, a Guatemalan exile, training in Honduras. Finally, on June 30, 1954, the Armas forces, supplied and trained by the United States government and under the direction of the United States Central Intelligence Agency(12, 13), overthrew the popularly elected Arbenz government. But this correlation with the apparent success of the agrarian reform program and the losses incurred by United Fruit Company is only spurious. According to Mr. Dulles, the Guatemalan takeover (which at the time he denied was instigated by the United States) was just a response to "Communist agitators . . . [who] dominated the social security organization and ran the agrarian reform program"(11; p. 97).

After the 1954 coup, lands that had been expropriated under the Arbenz government were given back to the large landowners and the agrarian reform program was essentially abandoned. Once again, resources were scarce (at least for the peasants) and the Malthusians could again cry overpopulation.

If the above analysis is too political or too historical or too anything else for the neo-Malthusians, let us return to the simple abstract model of the 100,000-acre island with 100 people. Any realistic analysis cannot stop with the simplified division of available resources (in this case acreage) by the number of people. The concentration of the resource and the changes in that concentration that are dictated by the political and economic system must be considered. For example, in this overly simple situation of the 100,000-acre island, I asserted that each of the 100 people had 100,000/100 = 1,000 acres of land, well over the required 100 acres. But very few populations approach, even approximately, that kind of equal division of resources. For example, if we superimpose the Guatemalan example on our 100,000-acre island with 100 people, we find that, according to the 1950 census, the pattern of concentration in land holdings would make it such that 2 of the 100 people would each have 36,000 of the 100,000 acres, 10 people would each have 1,350 acres, and 88 people would each have 162 acres. Thirty years later (following the previous example), four people would each have 18,050 acres, 20 people would each have 675 acres, and 176 people would each have 81 acres, below the assumed requirement of 90 acres. For 88 percent of the population, the critical resource scarcity would come 120 years earlier than in the original egalitarian system!

Obviously, as long as the economic and political system generates an inequitable distribution of resources, even a nongrowing population will experience apparent resource scarcity well before resources are actually in short supply. Thus, it would seem that the very real resource scarcity currently felt by the majority of the people in the world should be analyzed in terms of the problems of distribution of resources rather than in terms of some imagined overpopulation problem. Indeed, numerous lines of evidence point to the overwhelming importance of distributional factors over simple considerations of population size. For example, in terms of energy and resource consumption, each United States citizen costs the world the equivalent of what somewhere between 25 and 500 Indians cost(14). Most of the world's resources are consumed by developed nations with low population growth rates. (The United States, with 5 percent of the world's population, alone consumes 42 percent of the world's aluminum, 28 percent of its iron, 63 percent of its natural gas, and 33 percent of its petroleum [15]. Of approximately 60 million metric tons of fish caught in 1967, over one-third was consumed by the developed nations, and 45 percent was consumed indirectly by them as fish meal fed to livestock. Only 8 million metric tons [14 percent] went to the hungry nations [16].) Country-by-country comparisons show, if anything, a lack of correlation between population density and hunger. (The population density of India is 164 people per square kilometer, while that of Great Britain is 228 people per square kilometer. Other interesting comparisons of numbers of people per square kilometer: Mexico 25—United States 42; Bolivia 4—West Germany 237; Ethiopia 20—Belguim 316; Pakistan [including Bangladesh] 118—Japan 277; Mozambique 9—Netherlands 315 [17]. Clearly, population density per se is not the cause of resource scarcity.) Generally, those countries with higher population densities have higher standards of living than those with lower population densities. It seems indisputable that present population density must be far less important than distribution of resources in generating resource scarcity. As Barry Commoner notes, population control "is like trying to save a sinking boat through lightening the load by throwing people overboard. One is constrained to ask if there isn't something radically wrong with the ship"(18; p. 254). (See also, references 19 and 20.)

Concerning the assumption of the inevitability of inequity, Hardin's arguments seem to be two: first, historically, it has been that way; second, no matter what we do, a few people will garner the lion's share of the wealth—if someone who has a lot gives his or her wealth away, someone else will take over that person's position and nothing will have changed. The historical argument is obviously vacuous and merits little comment. It is at least debatable whether historically it has always been that way. (One might argue that many primitive societies, such as the !Kung bushpeople, the Congo pygmys, or even the Australian aborigines, have egalitarian societies. [See "Sociobiology—A New Biological Determinism" in this volume.]) And, even if it always had been that way, that does not argue one way or the other for the future.

However, the second argument is curiously correct, within its self-defined limits. It is not a new argument by any means. According to Hardin (5; p. 562):

"I feel guilty about my good luck," say some. The reply to this is simple: Get out and yield your place to others. Such a selfless action might satisfy the conscience of those who are addicted to guilt but it would not change the ethics of the lifeboat. The needy person to whom a guilt-addict yields his [sic] place will not himself feel guilty about his sudden good luck. (If he did he would not climb aboard.) The net result of conscience-stricken people relinquishing their unjustly held positions is the elimination of their kind of conscience from the lifeboat.

Hardin seems to be reminding us of one of the basics of economic theory. Under capitalism, wealth and power will tend to concentrate in the hands of a small segment of the population. Hardin is merely reaffirming this principle, albeit somewhat superficially. Indeed, we might accurately paraphrase his words to read, "the net result of conscience-stricken capitalists relinquishing their unjustly accumulated capital to the labor force is the elimination of their kind of conscience from the capitalist class, and thus the world's economy."

The sad and frightening feature of this piece of the analysis is that most people approaching these problems from the point of view of ecology do not seem to see that inherent features of the capitalist system are the major driving force of those ever-present inequities. Hardin does not admit that his own analysis says, in part, that any population organized under capitalism will be driven to the lifeboat ethic, regardless of its size or its rate of growth.

Many would, I suspect, grant this—that capitalism has led to the lifeboat ethics in the first place. (They would probably also insist that any industrially organized society would generate the same consequences—"socialism too.") They would also probably admit that in generating the inequalities that presently exist, the developing world has been choked by the developed world, that the rich countries have been massively stealing the resources, squandering the labor, and stunting the economic and political development of the underdeveloped countries. But the often heard counter is "I don't care about the past, I'm worried about the future." With that seemingly rational pragmatism, we would point out that we, personally, are not responsible for the present outrageous inequities and that as much as we deplore their existence we nevertheless feel that we must face up to their realities. That which exists now is what we must work with. From here on in, let us keep distribution constant. But what this fails to acknowledge is that it is impossible to keep distribution constant when the very force that has in the past determined distribution remains in effect. If we accept capitalism as the mode of economic organization, Hardin's second assumption is quite correct, inequities *are* inevitable. However, he is certainly not correct if we admit to a larger world view.

The third assumption is perhaps the most pernicious of all—that inequities are, in fact, desirable. This assumption is derived from an elitist conservation ethic. For example, consider the dawn redwood tree (one of Hardin's actual examples)(21). Long thought to have been driven to extinction by axe-wielding *Homo sapiens,* the species was rediscovered in isolated pockets in China. These pockets correspond to former or present locations of temple gardens. Hardin paints the historical picture of the peasant seeking firewood and the priest protecting the firewood (the dawn redwood tree). The conclusion is that without those priests—a privileged class—protecting those redwood trees from those peasants, the dawn redwood would indeed be an extinct species. Thus, if we are to preserve nature for posterity, some privileged class must be relied upon to do so, because common people are more concerned with common problems such as where to find food and shelter.

Again, we might argue that this is essentially correct; that as long as "common people" are a distinguishable part of society and are struggling for a less than adequate share of society's wealth, then, in order to preserve nature's beauties, we must have a privileged class to do so. But such an argument must ignore several causative factors. First, environmental degradation is generally caused by large industry, owned by that same privileged class that is supposed to save civilization and dignity. The lower classes have not polluted Lake Erie or putrified the air in Gary, Indiana. Second, the tendency to cause environmental damage, whether perpetrated by the upper or lower classes, is dictated by political institutions.

For example, the abandonment of the agrarian reform program in Guatemala was caused by the political changes imposed by the CIA. In place of that agrarian reform is a colonization program in which vast areas of pristine rain forest are being turned into farmland. The conservationist must realize that these forests are being destroyed unnecessarily, that their destruction has nothing to do with population pressure but rather is being caused by an imposed political institution that seeks to avoid meaningful agrarian reform.

The real choice faced by the conservationist is not between equality in social justice versus preservation of nature, as Hardinists would have us believe. Rather, it is between the preservation of small islands of nature in a sea of vulgar exploitation versus the creation of a system of human organization that would promote a harmonious existence both among human beings and between humans and nature. Do we wish to preserve patches of nature by way of luxury and exception, or do we wish to preserve nature itself for all people's benefit?

LIBERAL ANALYSIS

About the time the population bomb was being dropped, numerous ecologists pointed out some very elementary facts, facts that were conveniently ignored by most of the popularizers of the population problem. While enforced birth control was being pushed as a solution by the popularizers, a less vocal group of ecologists was pointing to the apparent social causes of high birth rates—poverty, lack of social justice, oppression of women, and so on. After numerous abortive attempts to disseminate birth control devices into underdeveloped regions with high birth rates, the quiet voice of the ecologists talking about social causes of demographic parameters is beginning to be heard.

For example, Barry Commoner has recently provided an excellent analysis of the relationship between population growth rates and economic development(22). He notes that, if one graphs GNP against population growth rates, one sees a negative relation between the two; those regions with high GNP have low growth rates and those regions with low GNP have high growth rates. He further shows that the only effective means of lowering the high rates of population growth is to let the underdeveloped regions develop, indeed to even help them to develop. When their GNP reaches the level of only a fraction of our own, their birth rates will decline; their population will be under control. Commoner says(22; p. 59):

What the problem calls for, I believe, is a process that now figures strongly in the thinking of the peoples of the Third World: a return of some of the world's wealth to the countries whose resources and peoples have borne so much of the burden of producing it—the developing nations. . . . There is no denying that this proposal would involve exceedingly difficult economic, social, and political problems, especially for the rich countries.

And there he leaves it. It is as though we have suddenly discovered that there is something wrong with a grossly inequitable distribution of wealth, and having discovered it, we now just legislate it away.

In summing up the problem, Commoner says, "I believe that if the root cause of the world population crisis is poverty, then to end it we must abolish poverty. And if the cause of poverty is the grossly unequal distribution of world's wealth, then to end poverty, and with it the population crisis, we must redistribute that wealth, among nations and within them"(22; p. 59). During the late sixties, many of us were led to believe that (a) poverty exists (among other things) and (b) to end it, we must abolish overpopulation. Now, we have the new liberal analysis: (a) overpopulation exists and (b) to end it, we must abolish poverty.

To those of us who have wished to end poverty all along this is a cruel irony. We are told that the "real" problem that causes poverty is overpopulation. Having worked in and been frustrated with programs designed to alleviate poverty, a significant fraction of the activists of the late sixties readily jumped on that bandwagon. Having firmly established in our heads that overpopulation is a truly significant social problem, it is easy to forget that we brought it up originally because of the problems that it created, such as poverty. Now that it is accepted as a problem in and of itself, we are told that we must solve the problem we originally were attacking so as to solve the "overpopulation problem." But we seem to forget how difficult that problem was originally. That Barry Commoner could say, in 1975, ". . . this proposal would involve exceedingly difficult . . . problems . . . for the rich countries" speaks to the naivety that the population panacea seems to have encouraged among ecologists. The socialists and communists of the thirties knew that it was a difficult problem. The guerilla forces in Vietnam knew that it was a difficult problem. The left in Chile knows that it is a difficult problem. But the population bombers have *just* discovered that we must end the maldistribution of wealth and that this will not be easy. We can look forward to yet further insights as they begin figuring out the ultimate causes of maldistribution.

Be that as it may, the new liberal analysis has two effects, one good and one bad. It focuses energies away from the purely physical attempts to lower birth rates (disseminating birth-control devices, forced sterilization, etc.). That is the good effect. However, it leaves population as the central issue in the ecological crisis. That is the bad effect.

ORIGINS OF THE POPULATION PROBLEM

Some people think that the discussion of the "population problem" originated with Paul Ehrlich or even with Tom Malthus, but the following was written very much earlier:

It might be supposed that there was more need of a fixed limit to the procreation of children than to the amount of property, so that no one should beget more than a certain number, and that this total should be fixed with reference to the chances of human life, viz. to the probability of some of the children not living to grow up and to the infertility of a certain number of marriages. The absence of all regulations, as is the case in the generality of States, will necessarily prove a cause of pauperism among the citizens, and pauperism is the parent of sedition and crime.

The quote is from Aristotle's *Politics*(23; p. 59), written sometime between 335 B.C. and 323 B.C. It is an interesting quote, since Aristotle's concept of the optimal size population is one in which all ". . . citizens should know each other's character" and the upper limit on the population of a state is ". . . not so large that it cannot be comprehended in a single view"(23; p. 176).

Since the basic reasoning is the same for any population size, one could easily find similar statements in any period of history. Or one could construct the argument *de novo* simply by assuming that the existing technology for distributing the most limited resources will not change or will change slowly in comparison with population growth. Again, the argument is essentially independent of what the population size actually is. Nevertheless, it should be noted that Malthus was not the first, nor the last.

But, rather than detail such ancient history, it is more to the point to examine the recent obsession with overpopulation. That obsession began, not in the late sixties as is popularly believed, but with an intense campaign in the early fifties.

In an article entitled "Why the Population Bomb Is a Rockefeller Baby"(24), Steve Weissman documents in detail the rise of the population movement. The Hugh Moore Fund, set up in 1944 by the chairman of the board of the Dixie Cup Corporation, published pamphlets "to call to the attention of the American Public the dangers inherent in the population explosion." In one of its early pamphlets, "The Population Bomb," quoted in Hansen(25; p. 8), we see:

Hundreds of millions of people in the world are hungry. In their desperation they are increasingly susceptible to Communist propaganda. . . . U.S. taxpayers cannot feed the world. . . . Today the population bomb threatens to create an explosion as dangerous as the explosion of the H bomb. No time is to be lost. The peril mounts daily. Our way of life, if not the actual existence of ourselves and our children, is at stake.

In 1952, John D. Rockefeller II set up and acted as president of the Population Council. Soon thereafter, the Ford Foundation, Carnegie Foundation, the Commonwealth and Community Funds, the Mott Trust, and the Mellons contributed substantially to the Council. In 1957, an *ad hoc* committee from the Population Council, the Rockefeller Fund, the Conservation Foundation (founded by Lawrence Rockefeller), and Planned Parenthood published *Population: An International Dilemma,* in which population growth was seen as a grave threat to political stability. In 1959, in the report of President Dwight Eisenhower's Committee to Study the Military Assistance Program, General William H. Draper II urged the president to include national population plans and additional research on population control in legislation for aid to development. Eisenhower rejected the proposal, and the investment banker Draper went on to head up Planned Parenthood's World Population Emergency Campaign. Along with the Rockefellers, Fords, and others, he continued to press for heavy governmental involvement in population control. By Earth Day 1970, the "overpopulation" problem was firmly entrenched in the public's mind as the principal cause of our present ecological crisis. And still today (1976) in a full-page ad in the *Smithsonian Magazine* by the Environmental Fund, endorsed by Paul Erlich (Stanford biologist), Garrett Hardin (University of California biologist), C. W. Cook (chairman, General Foods Corporation), W. G. Phillips (chairman, International Multifoods Corporation), and J. Paul Getty, among others, we are told that "we are being misled by those who say there is a serious food shortage. . . . The problem is too many people"(26).

Why, during the big buildup of the population crisis, did we hear so little from consumer groups, unions, and working-class people? Why was the banner carried by the rich, the industrialists, the American ruling class?

We can begin to understand the reasons by looking at the development of international capitalism as it existed during that time. The union movements of the late thirties had made significant economic gains that increased the cost of labor. Inflation kept profits high, but not as high as they could have been with cheap labor. Foreign countries, besides being major suppliers of raw materials, offered abundant non-unionized and cheap labor. Many industrial concerns moved much of their labor-intensive operations overseas. This generated changes in the gross demographic patterns of many foreign countries. Especially important was the virtually instantaneous creation of a massive urban proletariat as people migrated from country to city. But such migration was good for business—it kept labor costs low, as did any kind of population growth.

In the mid- to late forties, the concept of scientific management and mechanization developed both here and abroad(27). The efficient use of workers through time-efficiency methods and the replacement of much human-labor power with machines produced a large unemployed class in major foreign urban areas. An industrial reserve army is one thing, but masses of unemployed workers, with no obvious future prospects for employment, is quite another. Whereas the excess population in the Third World had created masses of cheap labor in the thirties and early forties, presently most industrial expansion includes much capital-intensive development. The excess population that originally had served as an industrial reserve army now must be viewed as a potential threat. What would all those people do? Would that mass of unemployed people get angry or frustrated? As F. Osborn, chairman of the Population Council's *Ad Hoc* Committee, said in 1958, ". . . unless a reduction of births is achieved within a few decades, the hopes of great but under-developed nations for better conditions of life may prove futile. . . . Such a tragic failure to achieve the higher levels of living that should be possible could only bring disillusion, confusion, and the danger of resort to desperate measures"(28; p. 95). Exactly who is afraid of those "desperate measures"? If the unemployed masses in the Third World can be convinced that their problem is too many people, rather than those foreign businessmen, maybe they will not take "desperate measures."

USING THE POPULATION BOMB

If the ecological problems that exist are, in fact, severe enough to threaten our very lives and if there are always structural reasons why these problems cannot be solved, it would seem to follow that the structure itself must be changed. In fact, in the early sixties, it seemed as though we were on the brink of a mass awakening to this consciousness. The repeated disclosures of environmental rape began awakening people to the realities of ecology. A growing awareness of the fundamentally political nature of the environmental crisis might soon have followed. But science saved the day again! Nothing is structurally wrong—there are just too many people. Latin Americans starve not because our coffee and bananas grow on their farmland but because there are too many Latin Americans. The putrid air in Gary, Indiana, does not result from the steel corporation profits but from too many people demanding steel. Every ecological problem can be tied to an overabundance of people.

Once again, the contradictions that had been emerging and challenging the fundamental nature of the sociopolitical system were defused by scientists. The problems have been shown to be "biologically determined," the subdiscipline "ecology" providing the needed scientific rationale. Just as woman's role is what it is because of her hormones and black people are poor because of their genes, the ecological crisis exists because populations normally grow at exponential rates. The ecological crisis is just as biologically determined and therefore by implication free from political analysis as are sex roles and economic classes. As Paul Colinvaux said in the concluding remarks at The Institute of Ecology (TIE) Founders' Conference, "There is one thing that we as ecologists can yet tell the people, . . . and this is the true ecological cause of these large populations. They come about because people have retained the Darwinian breeding strategy of their primitive ancestors"(29; p. 12).

Nevertheless, the overpopulation ideology is backfiring on the ruling class in the United States. Whereas the original notions of triage and its successor, lifeboat ethics, seemed to be the logical extension of the original population propagandists, the new liberal establishment is turning around and biting the hand that feeds it. While accepting the overpopulation problem as a problem, the liberals are now calling for reforms that look very much like the demands of Third-World revolutionaries.

What we have, it seems, are two lines of popular development in the ecology movement. First is the lifeboat line, which, if taken to its logical conclusion, must be described as the fascist line. Second, we have the more or less liberal line that suggests that population can only be truly controlled by social reform, and included in that social reform should be a direct attack on the ecological problems themselves, with the use of existing institutions. We must feed the hungry, clean up pollution, develop sustained-yield natural-resource exploitation, and insure social justice. All we need do is spend as much money on these things as we did on getting a man to the moon, and we can escape from our ecological dilemma. But watch out: the time is close to destruction. We must act in a hurry!

This is the conventional wisdom we hear from the academic ecological establishment (few openly espouse the fascist line). What has been the response to these calls for quick emergency action? Has there been a rash of effective legislation to clean up the air, to redistribute the wealth, to stop the exploitation of the Third World? In fact, there has been some legislation, but, whenever it is effective, we run up against the first law of ecology and capitalism: you must choose between a clean environment and a healthy economy, death by poisoning or by starvation.

So the response to liberals' calls for quick emergency action has not usually been the enactment of effective legislation. In fact, it appears that such legislation is virtually impossible under the present system. But there *has* been a response.

THE NEW ECOLOGICAL DETERMINISM

One of the basic and universally agreed upon results of generations of ecological research is that ecological systems are exceedingly complex; everything is connected to everything else. Indeed, this theme is echoed in two of the most popular major studies on global ecological problems. The first, sponsored by the Club of Rome and published under the title *The Limits to Growth*(30), emphasized the complexity of the system that is leading us to global ruin. They called for "the creation of a world forum where statesmen, policy makers and scientists can discuss the dangers and hopes for the future global system without the constraints of formal intergovernmental negotiations"(30; p. xi). The second, also sponsored by the Club of Rome and published as *Mankind at the Turning Point*(31), underscored the complexity of the system. This work also called for global planning for the development of a "practical" international framework: ". . . a number of meetings are already planned with public figures, political leaders of different parts of the world . . . [for the] development of a practical international framework . . . "(31; p. xi). Those attending such meetings are largely from the United States, Western Europe, and Japan.[1]

What will this "world forum" be like? What will constitute the "practical international framework"? The scientists who discovered the immense complexity of the world ecosystem now call for a "world forum" or an "international framework." Who can work effectively on such a large scale? What group could be well organized enough to even begin such a huge task? One group of people seem to think they have the answer. Carl H. Madden, chief economist of the United States Chamber of Commerce, said in 1973(33; p. 84):

The explosive growth . . . of the private multinational corporations is probably the single most important step taken to develop the technological means to bring adequacy to the world's people. . . . The multinational corporation . . . promises the most efficient use of world resources. . . .

Herbert C. Knortz, executive vice president of ITT, said in 1974(34; p. 535):

The multinational corporation constitutes one of the few institutions which effectively concern themselves with efficient utilization of natural resources. . . . In the case of the larger multinationals, operational effectiveness and cooperative relationships have been honed to a fine point of efficiency by decades of experience. . . . The multinationals represent the best hope at the present time for the advancement of a world economic community which will yield "most for the most"—greater benefit to the people of all countries.

And finally, in either an admission of guilt or an attempt to prophecy, James S. Kemper, Jr., president of Kemper Insurance Companies, said in 1973(35; p. 248):

Corporate managers are paid more money than any other large group in our society. We have more perquisites and privileges. We can experience the ultimate in job satisfactions. In a very real sense, we manage the future of the world, for as far ahead as anyone can see.

In their recent book *Global Reach,* Richard Barnet and Ronald Muller document the incredibly rapid growth of multinational corporations. In speaking of the ecologist's call for regional planning and management, they note the growing attitude of global corporations: "The world's only global planners argue that huge organizations alone are capable of planning on a planetary scale to solve the colossal problems now confronting mankind." Then, they point out what must be obvious to almost anyone who has thought about it, that ". . . the evidence is uncomfortably compelling that the kind of planning that hitherto has produced impressive corporate growth produces the wrong kind of social growth. The disparity between what modern man needs and what the modern corporation produces appears to be widening"(36; p. 359).

It is ironic that those who cause the disease insist that, if we contract just a little more of the disease, our symptoms will go away. From the incipient beginnings of competitive capitalism through the evolution of monopoly capitalism and its international ramification to the global reach of the multinational corporation, the various forms of the so-called free-enterprise system have plundered this planet and are the principal agents bringing us the now-popularized ecological crisis. Yet, they now want us to encourage their further development. And a new excuse is evolving. The science of ecology tells us we must plan on a global scale. The only ones truly experienced in successful planning on a global scale are the executives of the multinationals, etc., etc., etc.

[1]For an excellent review of these two books, see Jhirad, Lowe, and Strigini(32).

Scientists and technicians, in their sadly naive attempts at remaining "objective" in pursuing their "value-free" science, have frequently been made into unwilling conscripts, who unwittingly serve the very cause of the problems they wish to solve. Ecologists, I think, are the most obvious example of such service. In a truly passionate desire to save the world from ecological disaster, they have sometimes sacrificed a complete analysis of cause and effect for the pragmatic "what can we do here and now." If we cannot hope to control the monster United Brands, maybe we can talk Latin Americans into having fewer babies. If we do see the need for regional planning, let the global corporations do it, since no one else will in the near future.

This kind of copout, obviously, will not help to solve the problem; indeed, it will only exacerbate it. What we need to do is to attack the problem at its core. We must eliminate the fundamental system that holds profit above ecology, that continues to enslave both humanity and nature.

It all started with a simple idea. A long time ago, someone realized that it was possible to steal a piece of someone else's productivity. That little innovation influenced the development of human culture perhaps more than the lever, the wheel, the steam engine, or any other technological breakthrough. With the implementation of this idea, the fundamental relationships of human to human and human to nature changed. That change induced an evolving process that seems to be culminating today.

Yes, we are in a crisis situation. Fascist solutions are both humanistically and ecologically unsound. Liberal solutions cannot be achieved within our present socio-economic system. Clearly, the solution to environmental problems requires a radically different kind of politics.

References

1. *Time.* 1972. Test on taconite: Dumping of mine waste in Lake Superior. *Time* 99:44.
2. Huston, W. T. 1974. Corporate involvement in government. *Vital Speeches* 40:272-74.
3. Lerza, C. 1975. The world food conference: A summary. In *Food for the people, not for profit,* eds. C. Lerza and M. Jacobson. New York: Ballantine Books.
4. Paddock, W., and Paddock, P. 1967. *Famine—1975: America's decision—Who shall survive?* Boston: Little, Brown and Company.
5. Hardin, G. 1974. Living on a lifeboat. *BioScience* 24:561-68.
6. Hardin, G. 1971. Editorial. *Science* 172:1297.
7. *National Observer.* March 29, 1975, p. 1.
8. Vandermeer, J. 1976. Hardin's lifeboat adrift. *Science for the People* 8(1):16-19.
9. Stein, G. H. 1968. *Hitler.* Englewood Cliffs, N.J.: Prentice-Hall.
10. Maser, W. 1970. *Hitler's* Mein Kampf: *An analysis.* London: Faber and Faber.
11. Melville, T., and Melville, M. 1971. *Guatemala, another Vietnam?* Middlesex, England: Penguin Books.
12. Marchetti, V., and Marks, J. 1974. *The CIA and the cult of intelligence.* New York: Dell.
13. Jonas, S., and Tobis, D. 1974. *Guatemala.* New York: North American Congress on Latin America.
14. Davis, W. 1971. *Readings in human population ecology.* Englewood Cliffs, N.J.: Prentice-Hall.
15. U.S. Bureau of Mines. 1970. *Mineral facts and problems.* Washington, D.C.: U.S. Government Printing Office.
16. Borgstrom, G. 1970. The harvest of the seas: How fruitful and for whom. In *The environmental crisis,* ed. H. W. Helfrich, Jr. New Haven, Conn.: Yale University Press.
17. United Nations. 1969. *U.N. demographic yearbook, 1969.* New York: United Nations.
18. Commoner, B. 1971. *The closing circle.* New York: Bantam Books.
19. *Science for the People.* 1974. Not better lives, just fewer people. *Science for the People.* 6(1):18.
20. *Science for the People.* 1975. Science versus ethics. *Science for the People* 7(4):14-16.
21. Hardin, G. 1975. Address at the University of Michigan, Ann Arbor, March 26, 1975.
22. Commoner, B. 1975. How poverty breeds overpopulation (and not the other way around). *Ramparts* 31:21-25, 58-59.
23. Welldon, J. E. C. 1912. *The* Politics *of Aristotle.* London: Macmillan.

24. Weissman, S. 1970. Why the population bomb is a Rockefeller baby. In *Eco-catastrophe*, ed. the editors of Ramparts, pp. 26-41. San Francisco: Canfield Press.
25. Hansen, J. 1970. *The "population explosion," how socialists view it.* New York: Pathfinder Press.
26. *Smithsonian Magazine.* March 1976, pp. 28-29.
27. Braverman, H. 1974. *Labor and monopoly capital: The degradation of work in the twentieth century.* New York: Monthly Review Press.
28. Osborn, F. 1960. Population: An international dilemma. In *On population, three essays.* New York: New American Library.
29. Colinvaux, P. 1975. The coming climactic: Concluding remarks at The Institute of Ecology. *Bulletin of the Ecological Society of America* 56(4):11-14.
30. Meadows, D. H., Meadows, D. L., Randers, J., and Behrens, W. W. III. 1972. *The limits to growth.* New York: Universe Books.
31. Mesarovic, M., and Pestel, E. 1974. *Mankind at the turning point.* New York: E. P. Dutton and Company/Reader's Digest Press.
32. Jhirad, D., Lowe, M., and Strigini, P. 1975. The limits to capitalist growth. *Science for the People* 7(3):14-19, 34-37.
33. Madden, C. H. 1973. The multinational corporation, world money. *Vital Speeches* 40(3):84-87.
34. Knortz, H. C. 1974. The multinational corporation, an economic institution. *Vital Speeches* 40(17):535-40.
35. Kemper, J. S. 1973. Corporations: Power, ethics, and social responsibility. *Vital Speeches* 40(8):248-49.
36. Barnet, R. J., and Muller, R. E. 1974. *Global reach.* New York: Simon and Schuster.

Ecology, Society, and the Myth of Biological Determinism

Murray Bookchin

Despite the facile rhetoric that emphasizes our need to live in harmony with the natural world, a domineering attitude toward nature is still very much part of the prevailing sensibility. Our technology still rests on the assumption that we must control the natural world, and we almost unknowingly incorporate a Promethean attitude that reveals itself in countless subtle ways. A sinister reductionism persists in much of our thinking, a tendency to dissolve the qualitative features of both nature and society into quantitative ones, as though our interpretation of reality were neither precise nor meaningful (by which we normally mean "useful") unless it is stated in mathematical terms. We form our images of the world in terms of machines and arithmetical data, a body of images that we borrow from the technical sciences and statistical sociology. Human beings, in effect, become mere objects to be manipulated with an icy disregard for their freedom and autonomy.

To further subvert individual freedom and autonomy, we are now being told that our world is made up of institutions that are all but genetically emergent, indeed, institutions that stem directly from our morphology and physiology. What makes this notion so sinister, in my view, is that it tends to take existing conditions as they are and render them almost totally unchangeable. It would seem, according to this viewpoint, that we can only elaborate our "biological nature"—be it a nature based on human aggression or mutual aid—but we cannot hope to fundamentally change it, at least not without the aid of our friendly geneticists. One commonly encounters the analogy between human beings and fruit flies—that is, when we are not lectured on our vicious dispositions as primates. We multiply like fruit flies, parallel their population dynamics, and are subject to the material vicissitudes that determine their destiny. Given such presuppositions, it is not hard to draw the conclusion that our social institutions are cast largely in terms of racial background, kinship ties, age groups, and sex groups, and, from this framework, it is an easy step to add a biological foundation to our urban problems, wars, class conflicts, totalitarian institutions, and other social afflictions. In effect, we carry all our major social problems in our genetic material, and there is little we can really do about them other than endure them stoically.

Without denying a biological component to certain social problems, I would like to advance a very different point of view. I would like to contend that our attitude toward nature—including our attitude toward human nature and human biology—almost uncannily reflects the social attitudes we have toward each other. Accordingly, there have been as many different views toward the natural world as there have been different societies. Indeed, the projection of a given social system onto nature can be traced back through history into prehistory itself, at least so far as we can glean such projections from anthropology and mythology. We can see the evolution not only of different societies but of different attitudes toward nature that parallel social development itself.

Among many so-called primitive peoples, the natural world is seen somewhat as an egalitarian food web, closely reflecting the essentially egalitarian structure of the primitive community. Indeed, even the very institutions of the community are projected into the animal world, much as though human and animal "society" were completely congruent. Algonquin tribes that were organized in clan communities, for example, thought quite logically that beavers too were organized into clans that were presided over by clan elders. Our term *beaver lodge* still echoes a life-style in which American Indians of the Northeast lived in lodge-type structures. To cite almost randomly another example, albeit one closer to home: the dualism that was inhered in the classical Greek attitude toward nature patently reflected the dualism within Greek society itself, notably the split between Greek and barbarian, master and slave, male and female, the Olympian patriarchal deities and the chthonic matriarchal deities of an earlier civilization. The Greeks, in effect, had projected onto the natural world the very dualities that marked their social life. The medieval society that succeeded upon the classical world of antiquity established virtually the same feudal system in nature that we encounter in manorial society, a projection that goes so far that it extends to heaven and hell, as well as to nature. We still retain this medieval imagery when we describe certain animals as "kings of beasts" and others as "lowly ants," thereby reflecting a social world in which rule was the privilege of monarchs and aristocrats and subservience the more dubious privilege of "lowly" serfs and peasants.

With the emergence of a market economy, all the cooperative relations based on guild ties, not to speak of ties formed by clans and extended families, were virtually dissolved. Social life, whether communal or corporate, was totally atomized and the individual reduced to a mere monadic "buyer" and "seller," with each individual seeking the advantage of the other in a nexus of savage egotism and self-interest. Given this new commercial jungle, which we persist in misnaming a "society," it is not surprising to find that a new image of the natural world emerged. Nature, too, was seen as a jungle composed of avaricious, aggressive competitors, in short, of mere prey and predators. The social Darwinian description of nature, with its emphasis on the survival of the fittest and a claw-and-fang mode of natural selection, precisely reflected the relationships that prevailed in the nineteenth-century market place. The fit is almost perfect, and it is hard to say whether natural Darwinism produced social Darwinism or the very reverse. One has only to look at Malthus, whose work precedes the publication of *The Origin of Species,* to ask which is the parent of which. Today, as we move from the "free market" capitalism of the last century into an era of corporate state capitalism, an era marked by exaggerated planning, mass mobilization, and quantification, a new image of nature seems to be emerging—a cybernetic model in which quantity and quantification tend to yield a largely statistical interpretation of natural phenomena. Systems analysis, a paradigm so congenial to the corporate and bureaucratic mind, seems to be taking over where the social Darwinism of the "free" market place left off, a development that is reflected by our reverential use of such terms as *input, output, feedback, energy,* and so on, *ad nauseum.*

To sum up: the very notion that nature is an object to be dominated by humans stems, ironically enough, from the domination of human by human. Throughout history and prehistory, we have projected our social relationships onto the natural world so that it is quite impossible to speak meaningfully of "natural ecology" without also speaking of social ecology. By the same token, we cannot expect to have any biology in our conflict-ridden society that is so free of social interpretation that we can truly call it an "objective" science. Nor can we form a vision of nature that consists of "value-free," "brute" scientific facts, assuming that any "facts" could or should be "brutish" or free of any moral dimensions. In fact, we repeatedly project our social views (be they formed by the market economy, the patriarchal family, the class structure, or the ubiquitous system of domination), onto every aspect of nature, whereupon we, riddled with all our social preconceptions, return to the natural world to reinforce in society the very notions that were projected onto nature in the first place. This, I submit, is the authentic "feedback" principle that nourishes biological determinism and all its sinister conclusions for humanity.

Regrettably, the environmental movement, however well intentioned, has often fallen victim to this basic fallacy. The most compelling example is the population issue. I can hardly think of a more dramatic example of how we have adopted certain social views, projected them upon nature, and then permitted them to bounce off to reinforce the social views we initially established. Whatever the future may hold, I contend that today, and in the future, we have a "population problem" primarily in terms of our prevailing social parameters and goals. German fascism was thoroughly aware of the social basis of demographic theory and demographic projections, an awareness that entered into its formulation of racial and population policies. By the 1930s, as machinery began to replace labor and as large masses of unemployed people began to create a highly restive social situation, the issue of controlling the rebellious, technologically "superfluous" poor became an issue of focal concern. We do not have to go to Garrett Hardin's "lifeboat ethic" (see Vandermeer's article in this volume) to find the earliest evidence of a modern problem. Viewed from the standpoint of German fascism's imperial interest, Europe was "excessively overpopulated" even when population in the West was palpably declining and France was offering bonuses for large families to acquire cannon fodder for its armies. The attitude that fostered the Nazi atrocities of the Second World War stemmed from an image of population dynamics and, more fundamentally, from an image of nature that can be unerringly derived, in turn, from a distinct social order and distinct social needs.

Notwithstanding many advocates of population control, human beings do not multiply like fruit flies. Human birth rates are as inseparable from social conditions as human death rates, and the factors that impel human beings to reproduce are profoundly influenced by cultural factors, not merely biological ones. Alter the status of women in such a fashion that they are viewed not as mere sexual objects but as full human subjects who have a meaning and purpose in life, and the population growth rate can be expected to decline. Let a family enjoy a degree of material security that makes it possible not to view children as mere heirs or supporters of their elderly and infirm parents, and family sizes are likely to diminish. The sinister aspect of a largely biological approach to human birth rates is that, by treating human population dynamics as mere reflections of fruit fly population dynamics, social solutions to real or presumed demographic problems may be replaced by biological solutions—and the "ultimate" biological solution is genocide. It would have been hard to say this in the 1930s; it is much easier to say this since the world discovered Auschwitz. Unfortunately, well-intentioned liberals who so often initiate discussions in demography are not the ones to have the last word. The "practical" answers are usually furnished by reactionaries, and these answers are catastrophic in nature.

Allow me to note another aspect of the so-called "demographic problem" that confronts us today: if we were somehow to reduce the American population by a half without recourse to any authoritarian measures, I doubt that we would appreciably reduce the dimensions of our current ecological problems. Our society is, above all, a market society. It is a society created around "buyers" and "sellers." It lives by the basic maxim, "grow or die," a maxim that constitutes the very rationale of a market society. Despite the growing literature on the need to establish limits to growth, our society's very reason for existence is to produce for the sake of production and to consume for the sake of consumption. Entrepreneurs who do not produce will be swallowed up by their competitors. Accordingly, it would be easier to ask a green plant to desist from photosynthesis than to ask the bourgeois economy to desist from capital accumulation. Thus, I would contend that, if the population declined by a half in the United States, we would be asked to purchase three cars instead of two and to acquire a color television set for every corner of the room instead of one for each room. By the canons of a narrow biological deterministic approach, one might assume that a halving of the population would yield a halving of industrial output. A broader social outlook would reveal, indubitably with far more insight and accuracy, that the insane increase in production would continue as mindlessly as in the past and that the ecological crisis would increase in dimensions irrespective of population increases or decreases. My point is that a society based on production for the sake of production is inherently antiecological and would continue to devour the planet for reasons that depend very little on biological facts or population dynamics.

To create a new equilibrium between the natural and social worlds requires, in my view, a fundamental, indeed, revolutionary, reconstitution of society along ecological lines.

I would like to emphasize the word *ecological* in connection with my remarks. As we try to deal with the problems of an ecological society, the term *environmentalism* begins to fail us. Environmentalism tends increasingly to reflect an instrumentalistic sensibility in which nature is viewed merely as a passive habitat, an agglomeration of external objects and forces that must be made more serviceable for human use, irrespective of what these uses may be. Environmentalism, in effect, deals with "natural resources," "urban resources," and even "human resources." Environmentalism does not bring into question the underlying notions of the present society, namely, that humans must dominate nature; rather, it seeks to facilitate that domination by developing techniques for diminishing the hazards caused by domination. The very notion of domination itself is not brought into question.

Ecology, I would argue, advances a broader conception of nature and of humanity's relationship with the natural world. It sees the balance and integrity of the biosphere as an end in itself. The typically holistic concept of "unity in diversity," so common in the more reflective ecological writings, could be taken from Hegel's works. This is an intellectual convergence that I do not regard as accidental and that deserves serious exploration by contemporary neo-Hegelians.

Furthermore, ecology requires that humanity show a conscious respect for the spontaneity of the natural world, a world that is much too complex and variegated to be reduced to simple Galilean physicomechanical properties. Some systems ecologists notwithstanding, I would hold with Charles Elton's view that "the world's future has to be managed, but this management would not be like a game of chess . . . [but] more like steering a boat"(1; p. 151). In this new era of quantitative science, we may well find that we have lost hold of qualitative truths that cannot be reduced to statistics, that cannot be reduced to energy flow or to mere mathematicial equations. I wish to make a defense of qualitative science and of intuition and dialectical insight and argue that they have an authority equal to, and at times even superior to, that of the conventional mathematical paradigm. As one who has been deeply concerned with ecology, not just with environmentalism, I believe this world can well be undermined by simplifying it, not merely by polluting it.

The most important task we face today, if there is to be any nature (mathematical or qualitative), indeed, if there is to be any society, is to rid ourselves of the social lenses that cause us to view nature as a hierarchically organized "system," as a mere body of resources to be dominated and exploited. So-called primitive societies that are based on a simple sexual division of labor and that lack states and hierarchical institutions do not experience reality as we do through lenses, nor do they categorize phenomena in terms of "superior" and "inferior," "above" and "below." In the absence of inequality, these truly organic communities do not even have a word for "equality." As the late Dorothy Lee observed in *Freedom and Culture*(2; p. 40), a superb discussion of the "primitive" mind, ". . . equality exists in the very nature of things, not as a principle to be applied. In such societies, there is no attempt to achieve the goal of equality, and in fact there is no concept of equality. Often, there is no linguistic mechanism whatever for comparison. What we find is an absolute respect for man, for all individuals irrespective of age and sex."

The absence of coercive and domineering values in these cultures is perhaps best illustrated by the syntax of the Wintu Indians of California, a people Lee apparently studied at first hand. Terms commonly expressive of coercion in modern languages, she noted, are so arranged by the Wintu that they denote cooperative behavior. A Wintu mother, for example, does not "take" her baby into the shade; she "goes" with it into the shade. A chief does not "rule" his people; he "stands" with them. In any case, he is never more than their advisor and he lacks coercive power to enforce his views. The Wintu "never say, and in fact they cannot say, as we do, 'I have a sister,' or a 'son,' or 'husband,'" Lee observed. "*To live with* is the usual way in which they express what we call possession, and they use this term for everything they respect, so that a man will be said to live with his bow and arrows"(2; p. 8).

"To live with"—the phrase implies more than a deep sense of mutual respect and a high valuation of individual voluntarism; it also implies a profound sense of oneness between the individual and the group. The sense of unity within the group, in turn, extends by projection to the relationship of the community with the natural world. Psychologically, people in organic communities must believe that they exercise a greater influence on natural forces than is afforded by their relatively simple technology, an illusion they acquire by group rituals and magical procedures. Elaborate as these rituals and procedures may be, however, humanity's sense of dependence on the natural world, indeed, on its immediate environment, never entirely disappears. If this sense of dependence may generate abject fear or an equally abject reverence, there is also a point in the development of organic society where it may generate a sense of symbiosis, more properly, of mutualistic interdependence and cooperation, that tends to transcend raw feelings of terror and awe. Here, humans not only propitiate powerful forces or try to manipulate them; their ceremonies help (as they see it) in a creative sense to multiply food animals, to bring changes in season and weather, and to promote the fertility of crops. The organic community always has a natural dimension to it, but now the community is conceived to be part of the balance of nature—in short, a truly ecological community or *ecocommunity* peculiar to its ecosystem, with an active sense of participation in the overall environment and the cycles of nature.

This outlook becomes evident enough when we turn to accounts of ceremonies among people in organic communities. Many ceremonies and rituals are characterized not only by social functions, such as initiation rites, but also by ecological functions. Among the Hopi, for example, the major agricultural ceremonies have the roles of summoning forth the cycles of the cosmic order and of actualizing the solstices and the different stages in the growth of maize from germination to maturation. Although the order of the solstices and the stages in the growth of maize are known to be predetermined, human ceremonial involvement is integrally part of that predetermination. In contrast to strictly magical procedures, Hopi ceremonies assign a participatory rather than a manipulatory function to humans. People play a mutualistic role in natural cycles; they facilitate the workings of the cosmic order. Their ceremonies are part of a complex web of life that extends from the germination of maize to the arrival of the solstices. "Every aspect of nature, plants and rocks and animals, colors and cardinal directions and numbers and sex distinctions, the dead and the living, all have a cooperative share in the maintenance of the universal order," Lee observed. "Eventually, the effort of each individual, human or not, goes into this huge whole. And here, too, it is every aspect of a person which counts. The entire being of the Hopi individual affects the balance of nature; and as each individual develops his inner potential, so he enhances his participation, so does the entire universe become invigorated"(2; p. 21).

We must create an ecological society, one in which there is a mutualism between the social world and the natural world, as in Hopi society. The problem we face is not so much our understanding of nature as it is our understanding of each other. We must try to harmonize our own social relationships and respect the diversity within society itself, or at least its potential for diversification and its potential, ultimately, for liberation. We cannot continue to accept existing conditions and merely analyze them as though we were faced with a statistical problem in logistics, nor can we work with a methodology that assumes that what is given must be accepted and manipulated; rather, we must recover those ethical, humanistic, and qualitative values that will harmonize our relationship with the natural world. Fundamentally, this means that we must develop a society in which we live in a harmonized relationship with each other. An ecological society is ultimately one in which domination, hierarchy, and class have been removed from the human condition. Consider a food web and let me ask who are the "kingly" animals and who are the "lowly" ones. All of the components of a food web are interdependent. Similarly, in society, we must eliminate castes, hierarchies, and systems of domination and exploitation. As long as we have domination in society, we will try to dominate nature. As long as we have a market society in which production exists for the sake of production, we will reduce nature to natural resources, to mere objects of exploitation.

If we were to follow through on what it means to produce an ecological society—one that would be in balance with the natural world—we would have to work with certain assumptions. First, there cannot be an ecological society when people cannot directly control the society in which they live, when they cannot fully grasp and control the conditions of their social existence. What I am presenting here is no more than the Hellenic attitude that we have to think on a human scale so that we can begin to comprehend all the conditions of life around us. This scale implies the decentralization of our cities and the decentralization of our technology, both with the view toward making it possible for all of us to understand how we technologically interact with nature and to make it possible for us to control our technologies directly, to directly determine our social destiny. In addition, we must use more passive systems of technology, which can be best utilized on a decentralized basis, such as solar energy, wind power, methane digesters, and the like.

I am arguing, here, that decentralization does not mean that we will wantonly scatter population over the countryside in small isolated households or countercultural communes—vital as the latter may be—but rather that we will retain the urban tradition in the Hellenic meaning of the term, as a city that is comprehensible and manageable to those who inhabit it. Call it a new *polis* if you will, scaled to human dimension that, in Aristotle's famous phrase, can be comprehended by everyone "in a single view."

An ecological society would also presuppose the elimination of domination as such. To abolish domination would mean not only a "classless society," such as the Marxists would have us attain, but, perhaps more significantly, the elimination of domination in every aspect of life, notably the domination of the young by the old and of women by men, not only "man by man." Consider, for example, the Wintu Indians, who have a very different attitude toward what we today would rank as "superior" or "inferior." They have no notion of what we would call a "village idiot" or a lunatic. Everyone in their society has something to contribute to the group, however unique or peculiar the individual's traits. No one is ranked pyramidally as "above" or "below." Each is a member of the group as a whole. This sensibility is fundamentally different from our own. "Weakness" has one set of attributes, strength has another. Intelligence or quickness of mind may be one trait; wisdom may be another; craftsmanship may be a third; but none is "superior" to the other. From this standpoint, the Wintus lack the pyramidal ranking of experience that enters into our outlook; the world becomes a gestalt—a harmonious integration of many different individual features—each of which participates in a richly varied common fund, called the community.

The reductionism of the biological determinists validates the *status quo*. It accepts the given and fixes it eternally in the genetic material of humanity. The essential achievement of human beings, and ultimately their essential destiny, is their ability to transcend the biological, not with a view toward dominating it, but with a view toward adding a new dimension to it—the dimension we call "consciousness." There is a sense in which we have a destiny in the biological world. We have evolved from nature into a society in which people can associate with each other, not on the basis of blood ties or sexual ties, but, rather, on the basis of a genuine community of interest and consciousness. This evolution can give us a freedom unknown to other forms of life. With this freedom, we can construct a free society that can react with nature—not to simplify or to demolish the natural world, but rather to reconstruct it, to help it develop in the "light of reason," to promote variety and harmony within it. A new balance has to be struck between humanity and nature so that the liberation of society can lead to a "conscious" nature, one that will enrich the world of life and thereby create a more stable basis for a peaceful, harmonized natural and social world.

References

1. Elton, C. S. 1958. *The ecology of invasions by animals and plants.* New York: John Wiley and Sons.
2. Lee, D. D. 1959. *Freedom and culture.* Englewood Cliffs, N.J.: Prentice-Hall.

Sociobiology

The vision of human social organization as a reflection of elements in the animal world is not limited to Lorenz's and Ardrey's aggressive instincts nor to the reproductive capacity of fruit flies. On a somewhat grander scale, Herbert Spencer envisioned the development of human institutions by a process fully as competitive and "red in tooth and claw" as primitive Darwinism(1).

A modern form of social Darwinism is now upon us. Linking together a great deal of the previous work in biological determinism, the discipline now known as sociobiology seeks to unify much of biological determinism through a theoretical construct based on elementary concepts of evolution through natural selection. The role of women in our society, the aggressiveness of humans, and many other human qualities are seen as the results of the working out of the laws of natural selection.

The most talked about work in this new field is E. O. Wilson's *Sociobiology: The New Synthesis*(2). The massive press blitz that followed its publication (Wilson was interviewed on several talk shows and was provided with press space in publications including *People* [3] magazine, *House and Garden,* [4] and *The New York Times Sunday Magazine,* [5]) has led to the semipopularization of the "new" discipline. At a recent convention of the American Group Psychotherapists' Association, a panel discussion was organized entitled "Termites, Apes, and Man—Survival Value of Group Behavior." The abstract of the talks stated that "clinicians and students of human groups will increasingly be able to base their work on foundations rooted in zoology and primatology." In the November, 1975, issue of *American Biology Teacher,* P. R. Gastonguay argues for the establishment of sociobiology courses in high schools, suggesting as one reason that ". . . several crucial dilemmas faced by our present-day human society can be quantifiably related to analogous phenomena in other animal species"(6; p. 481). Indeed, an entire high school curriculum entitled "Exploring Human Nature" has been written by sociobiologists and is being used in schools in 27 states(7).

Clearly, the practitioners of today's sociobiology are undertaking a project much more wide ranging than other forms of biological determinism, comparable in extent to the social Darwinism of Spencer. The Sociobiology Study Group of Science for the

People argues that the new "science" is not a science at all but merely a reflection of a particular view of human institutions.

References

1. Peel, J. D. Y. 1971. *Herbert Spencer: Evolution of a sociologist.* New York: Basic Books.
2. Wilson, E. O. 1975. *Sociobiology: The new synthesis.* Cambridge, Mass.: Harvard University Press.
3. Wilson, E. O., and Jennes, G. 1975. Sociobiology is a new science with new ideas on why we sometimes behave like cavemen. *People* 4:68-71.
4. Wilson, E. O., and Seebohm, C. 1976. Getting back to nature—Our hope for the future. *House and Garden* 148:66.
5. Wilson, E. O. 1975. Human decency is animal. *The New York Times Magazine*, October 12, 1975, pp. 38-50.
6. Gastonguay, P. R. 1975. A sociobiology of man. *American Biology Teacher* 37:481-86.
7. DeVore, I., Goethals, G., and Trivers, R. 1973. *Exploring human nature.* Cambridge, Mass.: Educational Development Corporation.

Sociobiology—A New Biological Determinism

Sociobiology Study Group

Biological determinism attempts to show that the present states of human societies are the result of biological forces and the biological "nature" of the human species. Deterministic theories differ in detail, but all have a similar form. They begin by describing the particular model of society that corresponds to the author's socioeconomic prejudices. It is then asserted that the characteristics of that society are a necessary consequence of the biological nature of the human species. Therefore, present human social arrangements are either unchangeable or, when altered, will demand continued conscious social control because the changed conditions will be "unnatural." Moreover, such determinism provides a justification for the *status quo,* although some determinists dissociate themselves from the consequences of their arguments. The issue, however, is not the motivations of individual creators of deterministic theories but the ways in which those theories operate as powerful forms of legitimation of past and present social institutions. (See Bookchin's article in this volume.)

It is not surprising that the model of society that turns out to be "natural" bears a remarkable resemblance to the institutions of modern market society, since the theorists who produce these models are themselves privileged members of just such a society. Thus, we find that aggression, competition, extreme division of labor, the nuclear family, the domination of women by men, the defense of national territory, and individualism are over and over stated to be manifestations of "human nature."

The precise form of biological determinism has varied, depending upon the particular social institutions to be justified and upon the state of biological science at the time. In the nineteenth century, when genetics had not yet been invented and biological evolution was still controversial, zoologists like Louis Agassiz, an anti-evolutionist, tried to prove the biological inferiority of races and classes by the evidence of comparative anatomy and embryology. On the other hand, the social philosopher Herbert Spencer, who was an evolutionist, saw society as being moulded by the survival of the fittest through struggle among individuals. In the twentieth century, DNA has replaced gross anatomical features as the supposed determinant of human nature. Now, we are told that the *genes* of blacks or of members of the working class, rather than the shapes of their skulls, doom them to an inferior status in a species in which the status hierarchy is itself an inevitable result of the "human genotype."

The earlier forms of determinism have now been fairly well discredited. Thus, the contention of Jensen and Herrnstein that there is a high heritability for IQ and a genetic difference between races or between social classes has been thoroughly debunked. Much of the evidence for high heritability has turned out to be either "cooked" or grossly biased(1). (See the article by Schwartz in this volume.) Moreover, it has been pointed out repeatedly that the developmental and environmental plasticity of a character are not related to the presence or absence of genes that influence the character. Further, the demonstration that some trait differs genetically between *individuals* does not give any information about the causes of differences between *groups* such as races and social classes(2).

The simplistic forms of the human nature argument given by Lorenz, Ardrey, Fox and Tiger, and others have no scientific credit and have been scorned as works of "advocacy" by E. O. Wilson. His book, *Sociobiology: The New Synthesis*(3), is the manifesto of a new, more complex version of biological determinism. But Wilson's *Sociobiology* is no less a work of "advocacy" than its already rejected predecessors.

Wilson begins his book with a chapter entitled "The Morality of the Gene," in which he defines sociobiology as "the systematic study of the biological basis of all social behavior. For the present it focuses on animal societies. . . . But the discipline is also concerned with social behavior of early man and the adaptive features of organization in the more primitive contemporary human societies"(3; p. 4). The book as a whole is intended to "codify sociobiology into a branch of evolutionary biology," encompassing all human societies, ancient and modern, preliterate and postindustrial, since he believes that "sociology and the other social sciences, as well as the humanities, are the *last branches of biology* waiting to be included in the Modern Synthesis"(3; p. 4). [Our emphasis.] Like his predecessors, Wilson examines sociobiological data and theories in the light of ethical philosophy, with the objective of understanding the human situation with enlightened scientific thought.

This is not a mere academic exercise. For more than a century, the idea that human social behavior is determined by evolutionary imperatives and constrained by innate or inherited predispositions has been advanced as an ostensible justification for particular social policies. Deterministic theories have been seized upon and widely entertained not so much for their alleged correspondence to reality but for their more obvious political value, their value as a kind of social excuse for what exists. A few examples illustrate the point.

Eight years before Darwin published the *Origin of Species,* Herbert Spencer, a nineteenth-century apostle of determinism, argued in *Social Statics*(4) that it was unnatural to attempt the eradication of poverty by social welfare schemes since "the poverty of the incapable, the distresses that come upon the imprudent, the starvations of the idle . . . are the decrees of a large, far-seeing benevolence . . . under the natural order of things society is constantly excreting its unhealthy, imbecile, slow, vacillating, faithless members . . ."(4; p. 353).

As early as 1900, some leaders of industrial corporations understood that there was something to be gained from the biologizing of social institutions. In that year, the head of the immense armaments firm of Alfred Krupp, later one of the principal supporters of Hitler and a principal beneficiary of his slave-labor camps, offered a lucrative cash prize to the winner of an essay contest on the question, "What does the theory of descendence teach us in regard to the internal political development and legislation of states" (quoted in Zmarzlick [5; 457-58]). Somewhat later, John D. Rockefeller, an admirer of Spencer, told a Sunday school class that "the growth of a larger business is merely survival of the fittest . . . the working out of a law of nature and a law of God" (quoted in Hofstadter [6; p. 45]).

Ethologists and other determinists have not always left it to others to draw political conclusions from their works. Consider the case of the Nobel Prize winning ethologist, Konrad Lorenz. Perhaps "those who have been charmed by his film of himself leading, like a mother goose, a brood of greylag geese about the farm will recoil from identifying his works as political"(7; p. 124). Yet, writing in Germany, in 1940, when the slave-labor program and other genocidal policies were already in effect, Lorenz explained that the traditional ethical philosophy of modern civilization had created biological problems comparable to those in the domestication of animals. In the process of civilization, we have lost certain "innate releasing mechanisms" that are normally retained to maintain purity of breeding stocks: "The selection for toughness, heroism, social utility . . . must be accomplished by some human institution if mankind, in default of selective factors, is not to be ruined by domestication-induced degeneracy. *The racial ideal as the basis of the state has already accomplished much in this respect*"(8; p. 71). "We must—and may safely—rely on the healthy intuition of the best elements among us and entrust to them the choices that will determine whether the nation flourishes or deteriorates. If this selection fails, *and the defective elements are not eliminated,* they will permeate our nation . . . just like the cells of a malignant growth . . ."(8; p. 75). [Our emphasis throughout.]

OUTLINE OF THE THEORY

In order to make their case, determinists construct a selective picture of human history, ethnography, and social relations. They misuse the basic concepts and facts of genetics, evolutionary theory, and ethology to assert things to be true that are unknown, ignoring whole aspects of the evolutionary process and stating that conclusions follow from premises when they do not. They then invent *ad hoc* hypotheses to take care of the contradictions and rely on a form of "scientific reasoning" that leads to conclusions that are untestable.

Like other deterministic theories, sociobiology begins by attempting to establish the existence of a syndrome of social and individual behaviors that is characteristic or universal in the human species. However, it is often unclear to what extent a particular trait is claimed to be universal. So, for example, in Wilson's *Sociobiology,* we are told that ". . . true spite is commonplace in human societies" (p. 119) and ". . . that man would rather believe than know" (p. 561), and that "human beings are absurdly easy to indoctrinate" (p. 562). These assertions leave unclear how much variation in behavior is implied by the words "common," "rather," and "absurdly easy." These are terms that arise from a world view in which individual human beings and societies are seen as manifesting, to different degrees, their essential "natures" and "tendencies." It is no accident that the description of this underlying essential nature bears a remarkable resemblance to the society inhabited by the theorists themselves. In Wilson's case, it is the modern market-industrial-entrepreneurial society of the United States.

Having established a picture of human nature that conforms to their immediate world, sociobiologists try to show that this has arisen in the course of human evolution through natural selection. To do so, they carry out a four-step theoretical argument. First, they try to establish that "human nature" and social organization are only special examples of more general categories of behavior and social organization in the animal kingdom. In this way, they try to establish an evolutionary history for these social phenomena. The biological continuity is sometimes asserted to be a direct genetic continuity between humans and other primates through common ancestry (*evolutionary homology*) and sometimes to be the result of common forces of natural selection that operate on different genetic bases in more distantly related organisms and produce evolutionary convergence (*evolutionary analogy*). Both arguments depend, as we will see, on an anthropomorphic treatment of the behavior of other animals.

By necessity, the second step is the assertion that the different aspects of human social behavior are, to various degrees, genetically programmed. This step is essential because the modern theory of evolution by natural selection demands that the evolving trait have a genetic base. Without genetic variation, there can be no evolution in a *biological* sense (although there may be in a cultural sense). Sociobiologists must be equally opposed to the psychological determinism of Skinner and to any theories of human psychosocial freedom that involve an autonomous development of human history according to its own dynamic.

In fact, no evidence is offered by sociobiologists for a genetic basis of human social behavior beyond arbitrary postulations of genes and the obvious fact that people and apes are genetically different. The chief error of this part of the theory is the supposition that the arbitrary elements into which human behavior has been partitioned by natural and social scientists (elements like "xenophobia," "territoriality," "spite," "conformity," "religion," and so on) are based on underlying elements of heredity. It is the classic error of reification by which mental constructs are treated as if they represented concrete, objective realities.

Third, sociobiologists must argue that each element of "human nature" is in itself adaptive, that is, that it leads to higher reproductive fitness. For each characteristic, such as love, spite, entrepreneurship, xenophobia, and so on, they must invent a plausible story to explain why possessors of the trait might leave more offspring in the remote future.[1] It is not necessary to show that the individual will leave more immediate progeny, provided one can show that the individual's relatives are sufficiently benefited (*kin selection*). Alternatively, an "altruistic" act toward an unrelated person might elicit a future response of benefit to the actor. As we will show, such wide latitude is left for imaginative invention that no conceivable observation could be at variance with this theory. A little ingenuity will always lead to an adaptive explanation. A theory with this much "give" is worthless, because it is nonfalsifiable.

Finally, sociobiology must cope with the fact that human societies differ very much from each other in their cultural form and content and that, despite the predictions and observations of population genetics of the slow rates of *genetic* change and the small *genetic* differentiation among human populations, many societies have changed very rapidly. However, the possibility of using population genetics as a quantitative test of the theory is closed off by the *ad hoc* introduction of a "multiplier effect"(3; pp. 11-13) that allows sociobiologists to equate any observed amount of cultural differences arbitrarily with any postulated degree of genetic change. What follows is a detailed examination of each of these elements of sociobiological theory, especially as elaborated in E. O. Wilson's *Sociobiology*.

Like other works of scientific advocacy, *Sociobiology* cannot be read as a consistent or rigorous exposition. There are numerous direct contradictions on important points, and sometimes specific assertions contradict the general arguments of the work as a whole. For example, in one place Wilson says that "genes have given away most of their sovereignity"(p. 550) and, in another, that "it does not matter

[1]Strictly speaking, evolutionary theory requires that the *genes* of those possessing the trait must be more represented in future generations.

whether aggression is wholly innate or acquired partly or wholly by learning" (p. 255). But, if taken at face value, these statements are completely at variance with the central point of *Sociobiology:* that natural selection has established genes in the human species that are responsible for the human social condition. Such contradictions enable the author, when criticized, to claim to have disproved the critics by choosing the appropriate statements. This is another example of the "plasticity" that has been built into Wilson's arguments.

A VERSION OF HUMAN NATURE

Since *Sociobiology* is an attempt to explain the detailed structure of human societies as the result of biological evolutionary adaptation through natural selection, its first task is to delineate the model of human nature that it needs to explain. While the model is sometimes explicitly stated (for example, *territoriality* and *tribalism* are explicitly stated [pp. 564-65] to be human universals), more often its elements are embedded in more general discussion.

Among Wilson's various aspects of human nature are:

1. Indoctrinability—"Human beings are absurdly easy to indoctrinate—they seek it"(p. 562).
2. Spite and family chauvinism—"True spite is commonplace in human societies, undoubtedly because human beings are keenly aware of their own blood lines and have the intelligence to plot intrigue" (p. 119). (Undoubtedly? See Harris [9; pp. 369-91].)
3. The nuclear family, with male dominance and superiority—"The building block of nearly all human societies is the nuclear family. . . . Sexual bonds are carefully contracted . . . and are intended to be permanent . . ." (p. 553). Among "general social traits in human beings" are listed "aggressive dominance systems, with males dominant over females" (p. 552).
4. Reciprocal altruism (as opposed to true unselfishness)—"Human behavior abounds with reciprocal altruism," as, for example, "aggressively moralistic behavior," "self-righteousness, gratitude, and sympathy" (p. 120).
5. Blind faith—"Men would rather believe than know" (p. 120).
6. Genocide (p. 575) and warfare (p. 572)—"The most distinctive human qualities" emerged during the "autocatalytic phase of social evolution" that occurred through "intertribal warfare" and through "genocide" and "genosorption." While social evolution may no longer occur in this fashion, we are prisoners of our own evolutionary history. "Some of the functions are almost certainly obsolete, being directed toward such Pleistocene exigencies as hunting and gathering and intertribal warfare" (p. 575).
7. Xenophobia (fear of strangers)—"Part of man's problem is that his intergroup responses are still crude and primitive, and inadequate for the extended territorial relationship that civilization has thrust upon him. . . . Xenophobia becomes a political virtue" (p. 656).

This list is not exhaustive and is meant only to show how myopically "human nature" is viewed, as though its essentials were manifested in modern Euro-American culture as described by orthodox social science.

To construct such a view of human nature, Wilson either must divorce himself from any historical or ethnographic perspective or engage in bad ethnographic and bad historical reporting. His discussion of the economy of scarcity is an excellent example. An economy of scarcity and unequal distribution of rewards is described as an aspect of human nature, as follows: "The members of human societies sometimes cooperate closely in *insectan* fashion, but more frequently they compete for the limited resources allocated to their role sector. The best and most entrepreneurial of the role-actors usually gain a disproportionate share of the rewards, while the least successful are displaced to other, less desirable positions" (p. 554). [Our emphasis.]

Note that this description of modern market-economy society is universalized over time and space, consigning the unstated cooperative "exceptions" to an "insectan" form of behavior. It is rather reminiscent of the predetente rhetoric about "insectlike" Chinese millions, identically blue clad and mindlessly regimented into cooperative (and "unnatural") brigades. In fact, there exists a great deal of ethnographic description that completely contradicts this conception of social organization. There are scores of known "egalitarian" societies that do not have more or less desirable positions or inequalities of shares. (See, for example, the literature on Eskimos [9, 10, and 11].) Wilson's description of the economy of scarcity as universal ignores the following types of present and past societies.

1. Those *not* differentiated in any significant way by "role sectors."
2. Those *without* scarcities that are differentially induced by institutions for different subpopulations of the society (i.e., all members have equal access to all resources and, in many documented cases, have lived, apparently indefinitely into the past, *without* having approached the limits of their resources).
3. Those *not* differentiated by lower and higher ranks or strata (see Krader [12; pp. 29-42] or chapters 2 and 3 of Fried [13], which contrast egalitarian societies with rank and stratified societies).
4. Those *not* characterized by depressed or deprived access to societal rewards.

Most of these "egalitarian" societies enjoy a state of universal access to resources that has been labeled, by some, "affluent"(11). It has been well known for at least 100 years(14) that, in the "simpler" societies of the Old and New Worlds, people shared food and held possessions in common and worked only a few hours daily to meet their needs. The evidence from such societies is particularly devastating because it is upon these very groups that Wilson tries to build his hypothetical reconstruction of ancestral societies.

Wilson's conceptualization of gender roles[2] and male "dominance" serve as other examples of his attempt to create a human nature. *Sociobiology* conveys a powerful underlying statement that substantiates male supremacy in humans and other animals. The sexist message in the book is carried in subtle and implicit ways.

First, Wilson and other sociobiologists perpetuate the common sexist usages of the English language. In *Sociobiology,* animal behavior is sometimes described in anthropomorphic terms that have strong sexist connotations. Two examples should here suffice: it is stated that courtship display (usually males toward females) "can be envisioned as a contest between *salesmanship* and *sales resistance*" and that there "will be a strong tendency for the courted sex to develop *coyness*"(p. 320). [Our emphasis.] Later, on the same page, we are told that "rampant *machismo* has evolved in some insects. . . ." Chapters 13 through 16, in particular, are replete with additional examples of this use of language. "Man" and "mankind" are frequently employed when Wilson is referring to humans, and his use of "he" and "his" as the universal pronouns further demonstrates the sexism throughout the book.

Second, Wilson follows the culturally pervasive tendency to emphasize or place higher value upon what is male. In *Sociobiology,* this is accomplished by describing female behavior after male behavior is described and by using more males in

[2]Oakley(15), Tresemer(16), and others have pointed out that it is more appropriate to use the term "gender" when speaking of social behavior and to use the word "sex" for biological characteristics.

illustrations. Such a subtle but effective organization reveals the male supremacist perspective of the book. In photographs and illustrations where only one sex is shown, it is rarely a female; where both sexes are recognizable, usually the females are peripheral in placement and activity—the male is outstanding. Explicit captions leave no doubt of the secondary status of the female.

Third, the sweeping generalizations about human behavior, which are pivotal to Wilson's conclusions, have been made with few references to the work of anthropologists, ethnographers, and sociologists. Such work is often full of sexist assumptions, since systematic bias occurs when, for example, the ethnographer describes women by using information presented by the men in a society or assumes that the men's view of the value of women's activities is shared by all members of the society. Wilson draws upon studies that suffer from such sexist bias without criticizing them. Another striking point is the paucity of ethnographic data he utilized. To bolster his analysis of male dominance, Wilson cites Fox(17), so that Wilson's "evidence" is an interpretation of Fox's book, which itself is an interpretation of selected ethnographic reports. Interestingly, Tiger and Fox(18) were dismissed (p. 551) for being "inefficient and misleading" in their analysis of human social behavior and for using the "advocacy method" in reconstructing human social evolution (p. 28).

Indeed, it must be recognized that sexism, the *socially prescribed* power men have over women, does exist in most societies. Such a cultural and political arrangement, which varies from one society to the next(15, 19, and 20), is quite different from the "dominance hierarchy" that Wilson applies to many animal groups.

Wilson's view of aggression and warfare provides yet another example of his efforts to ascribe a general biological basis to social behaviors that prevail in Western societies. The historical and ethnographical data together indicate a steady increase in absolute and relative rates of killing in warfare as one moves toward large-scale, complex, and increasingly "modern" societies. "Primitive" warfare is rarely lethal to more than one, or at most a few, individuals in an episode of warfare and is virtually without significance genetically or demographically(21). Warfare in "simple" societies is often carried out in skirmishes and single combat. A wounding may be enough to call off the "war." Genocide was practically unknown until state-organized societies appeared in history (as far as can be made out from the archaeological and documentary records). Cave art of the Pleistocene seldom shows killing of persons with weapons, only hunting of animals(22; pp. 74-76; 23; pp. 80-82).

Moreover, the use of the present state of "primitive" people to reconstruct human history is highly suspect among anthropologists. Virtually all known primitives have lived in the coercive environments of state-organized societies characterized by warfare, imperial expansion, and imposition of exclusive ethnic definitions or tribute and taxes on external groups, the so-called tribes(13; pp. 170-74). Today, many anthropologists think that the characteristics of all presently known tribal peoples may be a function of the structural relationship to state-organized societies. Not a primordial "primitive" condition at all, "tribalism" is generated by the presence of the state, which has replaced prior household and cooperative locality groupings.

It is in the description of human nature that sociobiologists must invoke what is the first in a series of ploys designed to make the system of explanation proof against any possible test. Realizing that history and ethnography do not support the universality of his description of human nature, Wilson claims that the exceptions are "temporary" aberrations or deviations. Thus, although genocidal warfare is (assertedly) universal, "It is to be expected that some isolated cultures will escape the process for generations at a time, in effect reverting *temporarily* to what ethnographers classify as a pacific state" (p. 574). [Our emphasis.]

This is nonsense on several grounds. First, as was indicated earlier, most known "primitive" societies do not commit genocide in any meaningful contemporary usage of that term. Second, most ethnographers do not use classifications like "pacific states." Third, if a group, known to have committed genocide once, does not do so for generations, it is absurd to assert that the society is really essentially genocidal but is temporarily not manifesting its real nature.

Another related ploy is the claim that ethnographers and historians have been too narrow in their definitions and have not realized that apparently contradictory evidence is really confirmatory. Thus, "Anthropologists often discount territorial behavior as a general human attribute. This happens when the narrowest concept of the phenomenon is borrowed from zoology . . . it is necessary to define territory more broadly . . . animals respond to their neighbors in a highly variable manner. . . . The scale may run from open hostility . . . to oblique forms of advertisement or *no territorial behavior at all.*" [Our emphasis.] "If these qualifications are accepted, it is reasonable to conclude that territoriality is a general trait of hunter-gatherer societies" (pp. 564-65).

These are examples of a general method employed by sociobiologists in an advocacy procedure that guarantees that no observation can contradict the theory. First, they make definitions that *exclude nothing,* for example, territoriality. If a particular behavior appears to be absent, it is claimed that it is present in a different form. This technique has much in common with Freud's theory of "defense mechanisms," according to which a psychic drive may appear in a disguised form, even reversed (apparent love is really hate) or not at all, because it is "repressed." Second, Western conceptual categories are reified and given an objective and concrete reality that they do not have. Anyone who tries to give concepts like "religion," "entrepreneurship," and "warfare" precise technical meanings soon discovers that they are arbitrary and that the boundaries they reify do not exist in nature. The Martian zoologist invented in Wilson's Chapter 27 would be most unlikely to use any of the biased conceptual categories on which the argument of *Sociobiology* is built.

HUMANS AS ANIMALS—THE MEANING OF SIMILARITY

To support a biological explanation of human institutions, it is useful, but not necessary, for sociobiologists to claim an evolutionary relationship between the nature of human social institutions and "social" behavior in other animals. First, such a relationship would be indirect evidence that human social behavior has a genetic basis, and, second, it would make more plausible the argument that human social behavior has evolved under the influence of natural selection.

Evolutionary biologists make a distinction between two kinds of similarity: *homology* and *analogy. Homologous* features are retained from common ancestors by way of genes inherited from the common ancestors—the bones of a human arm and a bat's wing, for example. Obviously, sociobiologists would prefer to claim homology as the basis for similarity in behavior between humans and other animals, for then they would have a *prima facie* case for genetic determination. In some sections of *Sociobiology* (table 27-1, for example), Wilson attempts to do this by listing "universal" features of behavior in higher primates, including humans. But claimed external similarity between humans and our closest relatives (which are by no means extremely close to us) does not imply identity of cause. Reciprocal altruism may exist in higher primates and, for all we know, it may be genetically coded in them. But the same behavior in humans may be purely learned and widely spread among human cultures for the simplest of all reasons: successful and complex social structures demand cooperation for mutual benefit, and the enormous flexibility of the human brain permits this adaptive action as one facet of its vast range of potential behaviors.

Analogous features share similar forms and functions, but represent separate convergent evolutionary developments mediated by similar forces of natural selection, the wings of a bat and an insect, for example. More often than not, Wilson argues from analogy. Such arguments operate on shaky grounds. They can never be used to assert genetic similarity, but they can be used as a plausibility argument for natural selection of human behavior by the following logic: first, humans and some other animals perform a similar but only analogous behavior; second, since this behavior is adaptive, it would be favored by natural selection; and, third, natural selection has operated on different genes in the two species, but it produced the same response as an independent adaptation to similar environments.

The argument is not even worth considering unless the similarity is so precise that identical function cannot be reasonably denied, as in the classic case of evolutionary convergence, the eyes of vertebrates and octopuses. Here Wilson fails badly, for his favorite analogies arise by a twisted process of imposing human institutions upon animals by forced comparisons. Then, having imposed human traits upon animals by metaphor, he rederives the human institution as a special case of the more general phenomenon "discovered" in nature. In this way, human institutions suddenly become "natural" and are simply inheritances from an evolutionary past. (See Bookchin's article in this volume.)

A classic example, long antedating *Sociobiology,* is "slavery" in ants. "Slave-making" species capture the immature stages of "slave" species and bring them back to their own nests. When the captured workers hatch, they perform housekeeping tasks with no compulsion as though they were members of the captor species. Why is this "slave-making" instead of "domestication"? Human slavery involves, first, members of one's own species, second, continued compulsion, and, third, the slave as an article of exchange, a commodity engaged in the production of commodities. Slavery is a human economic institution in societies that produce an economic surplus, with the slave as a form of capital investment. It has nothing to do with ants, except by weak and meaningless analogy. Why, for that matter, do we speak of "queens" and "workers" with all the implications about the inevitability of social hierarchy? The so-called "queen" may be more a captive of the "workers" than their ruler. In many species, she is nothing more than a laying machine, bloated with eggs and forced by the "workers" to remain in one spot and to produce continually.

Wilson goes beyond ant hierarchies by expanding the realm of these mock analogies to find rudimentary forms of barter (p. 551), division of labor (p. 298), culture (p. 559), ritual (p. 560), and esthetics (p. 564) among animals. But, if we insist upon seeing animals in the mirror of our own social arrangements, we cannot fail to find any human institution we want among them. This is argument by analogy in its most inadmissible sense.

GENETIC BASES OF BEHAVIOR

We can dispense with the direct evidence for a genetic basis of various human social forms in a single word: *none.* The genetics of human social behavior is in a rudimentary state for both methodological and political reasons. Because we cannot reproduce particular human genotypes over and over or experimentally manipulate the environments of individuals or groups, it has been impossible to gather even the rudiments of information about how genes influence human behavior. That is the methodological problem. Because certain biological determinists have a strong desire to convince people that intelligence, criminality, and violence are mostly the fault of genetic differences, a great deal of bad data, incorrect analyses, and even outright misrepresentations have been propagated in recent years, obscuring what was already a muddy prospect. That is the political problem. There is no evidence that conforms to the elementary rules of experimental design for the existence of genetic differences between individuals or groups with respect to social phenomena. Nor is there any evidence that such traits as xenophobia, religion, ethics, social dominance, hierarchy formation, slave-making, and so on, are in any way specifically coded in the genes of human beings. And, indeed, Wilson offers no such evidence. As a result, he makes a series of confused and contradictory statements about what is an essential element in his argument. So, for example, he speaks of "genetically programmed sexual and parent-offspring conflict" (p. 563), yet there is the "considerable technical problem of distinguishing behavioral elements and combinations that emerge . . . independently of learning and those that are shaped at least to some extent by learning. . . . Where both processes contribute, their relative importance under natural conditions is extraordinarily difficult to estimate" (p. 159). That is putting the difficulty mildly. Elsewhere, the capacity to learn is stated to be genetic in the species, so that "it also does not matter whether aggression is wholly innate or acquired partly or wholly by learning" (p. 255). But it does matter. If all that is genetically programmed into humans is "that genes promoting flexibility in social behavior are strongly selected" (p. 549) and if "genes have given way most of their sovereignty" (p. 550), then biology and evolution can offer no insight except the most trivial one that the *possibility* of social behavior has evolved. But Wilson does not believe that genes have given up their sovereignty where it counts, since, in the next phrase, he states that genes "maintain a certain amount of influence in at least the behavioral qualities that underlie the variations between cultures." He has stated as a *fact* that genetic differences underlie variation between cultures, when no evidence at all exists for this assertion and when some considerable evidence is against it. The only citation of evidence is that "moderately high heritability has been documented in introversion-extroversion measures, personal tempo, psychomotor and sports activities, neuroticism, dominance, depression and the tendency toward certain forms of mental illness, such as schizophrenia" (p. 550). But, except for recent studies of schizophrenia(24), there is no evidence for the inheritance of any of these traits, nor is there any evidence of any kind for genetic differences between cultural groups in these traits. Indeed, we are not told why sports activities, neuroticism, and schizophrenia are "behavioral qualities that underlie the variations between cultures."

If we take Wilson at his word, that he believes genes to have some role, but not an exclusive one, in cultural variation, then there is a large gap in the theoretical structure. For, if genes are not everything, then the dynamics of the evolution of human culture must include not only the rules of Mendelian inheritance but some theoretical structure of cultural transmission as well. Yet, in a work that claims to lay the foundations of a theory of the evolution of society, there is no hint of a theory of cultural transmission or of the possible interaction of cultural and genetic transmission. It is not that there is no completely articulated theory of cultural transmission, it is that there is no consideration of the problem at all. On the contrary, the problem of cultural transmission is dismissed by characterizing Dobzhansky's position, "Culture is not inherited through the genes, it is acquired by learning from other human beings," as an "extreme orthodox environmentalism" (p. 550). But Wilson's *effective* position, all qualifying phrases to the contrary notwithstanding, is an extreme hereditarian one, since the only dynamic of transmission and evolution considered is natural selection operating on Mendelian genes.

Since Wilson can adduce no facts to support the genetic basis for human social behavior, he tries two tacks. First, he uses the suggestion of evolutionary homology between primates and humans to imply a genetic basis in our species. But, as we discussed earlier, the evidence for homology as opposed to analogy is very weak. More important, the phylogenetic method does not tell us whether the traits that are relatively constant from group to group, the so-called *conservative traits,* are those that are most likely to be under direct genetic control or whether it is the traits that vary from group to group, the *labile traits,* that are genetically determined. In fact, Wilson confuses the issue by claiming both. At first, he claims that conservative traits are most probably homologous, but then he hedges: "The possibility remains that some labile traits are homologous between man and, say, chimpanzee. And conversely, some traits, conservative throughout the rest of the primates, might nevertheless have changed during the origin of man" (p. 551). Then, he reverses the argument to "heuristically conjecture" that phenotypically *labile* traits are those that are most likely to differ from one human society to another on the basis of genetic differences. The phylogenetic method is thus reduced to uselessness, since both constancy and lability are to be taken as evidence for a genetic basis. We are not told by the sociobiologist what sorts of traits are *not* likely to be genetically based. We must agree with Wilson that "the comparative ethological approach does not in any way predict man's unique traits. It is a general rule of evolutionary studies that the direction of quantum jumps is not easily read by phylogenetic extrapolation" (p. 551).

The second main tack is to postulate genes right and left. The technique is a simple one. Start by saying "if there were a gene for spite" and then go on with a long discussion of the consequences, dropping the "if" along the way. There are hypothetical altruist genes, conformer genes, spite genes, learning genes, homosexuality genes, and so on. An instance of the technique is on pp. 554-55. "Dahlberg (1947) showed that *if* a single gene appears that is responsible for success and upward-shift in status. . . . Furthermore, there are *many* Dahlberg genes. . . ." [Our emphasis.] Or, again (p. 555), "The homosexual state itself results in inferior genetic fitness. . . . It remains to be said that if such genes really exist they are almost certainly incomplete in penetrance and expressivity." The notion of confirming the existence of behavior genes that only have effect in some unspecified proportion of individuals carrying them (*incomplete penetrance*) and with some unspecified range of manifestation when they do have effect (*variable expressivity*) is interesting. (A standing joke among human geneticists is that, when some trait has defied attempts to unravel its inheritance [if any], it is said to be "a dominant gene with incomplete penetrance and variable expressivity.")

Another example is on p. 562: "If *we assume* for argument that indoctrinability evolves . . ." and "societies containing higher frequencies of conformer genes replace those that disappear. . . ." [Our emphasis.] Or how about "innate moral pluralism" (pp. 563-64)? For other examples, see nearly any page of Trivers (25 or 26).

The vagueness and contradictions of the genetic arguments appear at every level. Are we to assume that the human species is genetically uniform for the genes producing "human nature"? If so, how are we ever to detect such genes? On the other hand, perhaps we are genetically variable. If so, what mechanisms maintain the variation? Despite an early chapter on the mathematics of population genetics, not a single result of this highly developed theory, which addresses the issue of how genetic variation is maintained, is applied by Wilson to his speculations on behavioral genes because of the vagueness of his genetic hypotheses. The mathematical theory is so much window dressing.

Finally, we do return again to the error of reification. Geneticists long ago abandoned the naive notion that there are genes for toes, genes for ankles, genes for the lower leg, genes for the kneecap, and so on. Yet, the sociobiologists break the totality of human social behavior into arbitrary units, call these elements "organs" of behavior, and postulate particular genes or gene complexes for each.

EVERYTHING IS ADAPTIVE

The next step in the sociobiological argument is to try to show that the hypothetical genetically programmed behavior organs have evolved by natural selection. Two assertions are made, the first explicit, and second implicit: first, since natural selection insures that any genes that confer higher reproductive success on an individual or its relatives will increase in number in succeeding generations, behaviors and social structures that we observe must exist because they are *or were* adaptive; second, a trait that increases an individual's status, power, and wealth also increases its reproductive success.

There is relatively little discussion of this second assertion by sociobiologists, although, if it is not true, their argument falls. If natural selection is to have established the behavioral characters that lead to status and wealth, it can do so *only* if those characteristics somehow lead to a higher reproductive success. Evidence referred to by Wilson (p. 288) states that among laboratory mice, *Anubis* baboons, and Yanamamo Indians, the "dominant" males impregnate a disproportionate share of females. In fact, in the ethnographic literature, there are numerous examples of groups whose political "leaders" do not have greater access to mates (e.g., the Yaruro or American society). In general, it is difficult to demonstrate a correlation of any of the sociobiologists' "adaptive" social behaviors to actual differential reproduction.

The first assertion, that all human behavior is *or has been* adaptive, is an outdated expression of Darwinian evolutionary theory, characteristic of Darwin's nineteenth-century defenders who felt it necessary to prove everything adaptive. Although modern evolutionary theory is cleansing itself of such panselectionism, sociobiologists maintain the old line. It is a deeply conservative politics, not an understanding of modern evolutionary theory, that leads one to see the wonderful operation of adaptation in every feature of the living world. There is no hint in Wilson that evolutionary geneticists are deeply divided on the question of how important adaptive processes are in manifest evolution. Most evolutionary geneticists will admit that some fraction (the argument is over how large a fraction) of evolutionary change has occurred either in the absence of natural selection and adaptation or as an interaction between adaptive and random processes.

Second, and more important, modern evolutionary theory gives a large place to the *indirect* effects of selection. That is, organs or traits that are not themselves under direct natural selection may change their size or shape as a consequence of developmental or physiological links to features that are under direct selection. So, for example, many populations of animals undergo severe oscillations in numbers, not because natural selection has favored oscillation, but because selection has favored concentrating high reproduction in one age class, and that, *incidentally and for purely dynamical reasons,* leads to population oscillation. Indirect selection is a strain in evolutionary thought that goes back to Julian Huxley's(27) concept of *allometry,* by which he sought to avoid much of the tortured logic required by naive theories of direct selection(28). Allometry is the differential rate of growth of organs so that larger animals do not have all of their organs proportionately larger. For example, among species of primates, teeth do not increase proportionately with increased body size. Thus, large primates have proportionately smaller teeth than small primates as a result of a purely developmental relationship. Many evolutionary trends have been explained in this way, and it seems at least reasonable that many features of human social organization, if not all, may be involved in a complex system of correlation with other changes (for example, increased plasticity of neurological response and cognitive capacity.)

Thus, the major assertion of sociobiologists that behaviors and social structures exist because of their superior adaptive values is only an assumption that is nowhere tested. In fact, no tests are even proposed or any criteria established to allow a distinction between the natural selection model, historical models, and religious beliefs, as explanations for the diversity of manifest behavior. We cannot distinguish the possibility that parent-offspring conflict is the result of natural selection for conflict genes (p. 341) from the possibility that parent-offspring conflict in humans is an historically conditioned outcome of certain forms of social and economic arrangement. It never occurs to Wilson to suggest a test of the alternative hypothesis that "conformity" has not been selected by differential reproduction (pp. 561-62); instead, he claims that conformism and indoctrination result from an application of the ideology expressed by Daniel Webster when he said, "Education is a wise and liberal policy, by which property and life and the peace of society may be secured."

When we examine carefully the manner in which sociobiology pretends to explain all behaviors as adaptive, it becomes obvious that the theory is so constructed that *no tests are possible.* There exists no imaginable situation that cannot be explained; it is *necessarily confirmed by every observation.* The mode of explanation involves three possible levels of the operation of natural selection: one, classical individual selection to account for obviously self-serving behaviors; two, kin selection to account for altruistic or submissive acts toward relatives; and, three, reciprocal altruism to account for altruistic behaviors directed toward unrelated persons. All that remains is to make up a "just-so" story of adaptation with the appropriate form of selection acting. For some characteristics, it is easy to invent a story. The "genes" for dominance, aggression, genocide, deception, and hypocrisy will "obviously" be advantageous at the individual level. For other behaviors, rather more ingenuity is required. For example, in a recent high school text produced by sociobiologists I. DeVore, G. Goethals, and R. Trivers(29), students are asked to give an evolutionary explanation of why children hate spinach, when adults like it. The answer: spinach contains a substance that makes calcium unavailable to the body; children need calcium for growing bones; adults no longer need much calcium and so can benefit from the other nutritional advantages of spinach; therefore, genes causing children to dislike spinach, but allowing adults to like it, will be favored by natural selection(29; p. 115).

A rather less trivial example is the attempt of sociobiologists to explain homosexuality (which they assume to be genetic without evidence). Homosexuality would seem to be at a reproductive disadvantage since "of course, homosexual men marry much less frequently and have far fewer children." (No account is taken of the high frequency of homosexual encounters among "heterosexual" men, as shown by Kinsey. Indeed, the whole typology of "homosexual" and "heterosexual" is in question. And what about "homosexual" women?) But a little imagination solves the problem: "The homosexual members of primitive societies may have functioned as helpers . . . [operating] with special efficiency in assisting close relatives" (p. 555). The genes would then be "sustained . . . by kin selection alone." Here we can see the postulate of kin selection, introduced by Hamilton (30), operating to save the day when direct selection fails. If you cannot think of a reason why a trait will confer direct fitness on its bearer, you need only imagine why it might help the *relatives* of the bearer. In terms of spreading one's genes, it is worthwhile to sacrifice your own life to save two sisters, eight first cousins, and so on, since you will then be propagating your genes indirectly.

Altruistic behaviors when directed toward nonrelatives can also be explained in a similar manner. Only one more imaginative mechanism is needed to rationalize such phenomena as friendship, morality, patriotism, and submissiveness even when the bonds do not involve relatives. The theory of reciprocal altruism(25) proposes that selection has operated so that risk taking and acts of kindness can be recognized and reciprocated and the net fitness of both participants is increased. It requires that the actor and the beneficiary be able to remember the act and calculate the costs and benefits of reciprocity.

The trouble with all these explanations is that they render the whole system unfalsifiable. Nothing is explained because *everything* is explained. When individuals are selfish, that is explained by simple individual selection. When, on the contrary, they are altruistic, it is kin selection or reciprocal altruism. When sexual identities are unambiguous, it is because individual fertility is increased. When, however, homosexuality is common, it is the result of kin selection. The sociobiologists give us no example that might conceivably contradict their scheme of perfect adaptation. If one should arise, we will be told by Wilson that it is a "temporary reversion," as in those tribes who are in "what demographers describe as a pacific state" (p. 574).

There do exist data supporting the idea that kin selection is effective for some traits in social insects, just as there are large quantities of data that show the existence of individual selection for many traits in many organisms. It is not the existence of kin selection nor individual selection as phenomena that is at issue. It is, rather, the arbitrary invocation of these effects, together with the untested story about "reciprocal altruism," that makes the total system of explanation for all human behavior inaccessible to any test or possibility of falsification.

One of the major consequences of the creation of an infinitely flexible and vague system of explanation is that a general atmosphere of easy but confused rationalization for any observation is created. Kin selection, reciprocal altruism, and the superior survivorship of whole groups (group selection) become mixed up in an attractive and vaguely plausible scheme to explain anything. Thus, E. F. Foulks explains the existence of schizophrenia(31): prophets and shamans (Joan of Arc and the Seneca prophet Handsome Lake are given as examples) frequently hallucinate, hear voices, interfere in others' thoughts, and are presumed to be schizophrenics. These people help their societies cope with unusual stresses by inducing or leading them to new social and political institutions, and the society in turn rewards them with high status. In this way, schizophrenics may have evolved to help "effect social change when traditional methods fail." (Joan's high status seems not to have been correlated with high fecundity.)

This sort of fantasy is likely to become a popular intellectual pastime in the near future. We may even envision a parlor game, "Find the Adaptation."

VARIATIONS OF CULTURE IN TIME AND SPACE

We have pointed out that adaptive evolutionary arguments can be rendered totally untestable by postulating genes wherever the theory requires them and by telling "just-so" stories about adaptation. There does exist, however, one possibility of tests of such hypotheses, where they make specific *quantitative* predictions about rates of change of characters in time and about the degree of differentiation between populations of a species. Population genetics, the analytical base on which evolutionary explanations are built, makes specific predictions about such rates, based upon estimates of mutation rates, population sizes, selection intensities, and migration patterns. In addition, there are hard data on the degree of genetic differentiation between human populations for biochemical traits(32). Both the allowable rates of *genetic* change in time, from theory, and the observed *genetic* differentiation between populations are too small to agree with the very rapid changes that have occurred in human *cultures* historically and the very large *cultural* differences observed among contemporaneous populations.

To take a particular case, we can consider the rise and fall of Islam. In the sixth century A.D., the Arab world was made up of poor, pastoral people. In A.D. 622, Mohammed appeared, and, within 200 years, Islam was the supreme power in the West, both politically and culturally. Music, poetry, mathematics, science, historiography, all flowed in profusion from Baghdad and Cordoba, the two great capitals of power and culture in the tenth century. But, by 1492, the last Moors had been ejected from Spain and the Arab people were again a poor pastoral culture. This cycle, occurring over fewer than 30 generations, was too rapid by orders of magnitude for any large change by natural selection. It would appear that really significant cultural changes occur by some nongenetic means and that an understanding of human history and fortunes must be sought where it has always been, not in genes but in culture.

The same problem arises from the immense cultural differences between contemporaneous groups, since we know from studies of enzyme-specifying genes that at least 85 percent of that kind of human variation lies *within* any local population or nation, with a maximum of about 8 percent between nations and 7 percent between major races(33).

Wilson acknowledges and deals with both of these dilemmas with one bold stroke—he invents a new phenomenon. It is called the "multiplier effect" (pp. 11-13 and pp. 569-72). The multiplier effect postulates that very small differences in the frequencies of genotypes will result in major cultural differences. Thus, only a small increase in the frequency of "superior" types among the Arabs might result in immense changes in their political power and cultural riches. In like manner, very small differences in the frequency of hypothetical genes for altruism, conformity, indoctrinability, role playing, and so on, could move a whole society from one cultural pattern to another, even though most of the people were not different biologically. In a discussion of the multiplier effect (pp. 11-13), Wilson contrasts the large differences in behavior between closely related species of insects and of baboons, with supposed small differences in genes between the species. There is, however, no evidence about the amount of genetic difference between these closely related species, nor do we know how many tens or hundreds of thousands of generations separate the members of these species pairs since their divergence. Thus, we cannot say whether the behavioral differences are large or small relative to the genetic divergence or to the time of separation of these organisms.

There is the added problem of why nonhuman animals do not show equally rapid evolution and equally dramatic interpopulation variation in "social" traits. This anomaly is disposed of by yet another *ad hoc* hypothesis, the "threshold effect" (p. 573), according to which organisms must reach a certain (unspecified) level of social complexity before the multiplier effect will operate. Thus, some of the "'noblest' traits of mankind including team play, altruism, patriotism, bravery on the field of battle" may be the "genetic product of warfare," but the threshold effect prevents their development in lower animals (p. 573). But the multiplier effect, by which any arbitrary but unknown genetic difference can be converted into any cultural difference you please, and the threshold effect, by which any arbitrary level of social development can be postulated to account for any failure of generality, are pure inventions of convenience without any evidence to support them. They have been created out of whole cloth to seal off the last loophole through which the theory might have been tested against the real world.

AN ALTERNATIVE VIEW

It is often stated by biological determinists that those who oppose them are "environmental determinists" in the manner of Skinnerian psychology who believe that the behavior of individuals is precisely determined by some sequence of environmental events in childhood and youth. But such an assertion misses the point and reveals the essential narrowness of viewpoint in such deterministic ideologies. First, these ideologies see the *individual* as the basic element of behavior and determination, while society is simply the sum of all the individuals in it. For biological determinists, social institutions result from natural selection operating on individuals either directly or through their kin. But the truth is that the individual's social activity is to be understood only by first understanding social institutions. We cannot understand what it is to be a slave, or a slave owner, without first understanding the *institution* of slavery, an institution that defines both slave and owner.

Second, determinists assert that the evolution of societies is the result of changes in the frequencies of different sorts of individuals within them. But this confuses cause and effect. Societies evolve because social and economic activity alter the physical and social conditions in which these activities occur. Unique historical events, actions of some individuals, and the alteration of consciousness of masses of people interact with social and economic forces to influence the timing, form, and even the possibility of particular changes, but the individuals are not totally autonomous units whose individual qualities determine the direction of social evolution. Feudal society did not pass away because some autonomous force increased the frequency of entrepreneurs. On the contrary, the economic activity of Western feudal society itself resulted in a change in economic relations that made serfs into peasants and then into landless industrial workers, with all the immense changes in social institutions that were the result.

Finally, determinists assert that the possibility of change in social institutions is limited by the biological constraints on individuals. But we know of no relevent constraints placed on social processes by human biology. There is no evidence from ethnography, archaeology, or history that would enable us to circumscribe the limits of possible human social organization. What history and ethnography do provide us are the materials for building a theory that will itself be an instrument of social change.

References

1. Kamin, L. 1974. *The science and politics of I.Q.* New York: Halsted Press.
2. Lewontin, R. C. 1970. Race and intelligence. *Bulletin of Atomic Scientists* 26:2-8.
3. Wilson, E. O. 1975. *Sociobiology: The new synthesis.* Cambridge, Mass.: Harvard University Press.
4. Spencer, H. 1851. *Social statics: The conditions essential to human happiness specified, and the first of them developed.* London: Chapman.
5. Zmarzlick, H. G. 1973. Social Darwinism in Germany viewed as an historical problem. In *From republic to reich: The making of the Nazi revolution,* ed. H. Holborn. New York: Vintage Books.
6. Hofstadter, R. 1955. *Social Darwinism in American thought.* Boston: Beacon Press.
7. Eisenberg, L. 1972. The human nature of human nature. *Science* 176:123-28.
8. Lorenz, L. 1940. Durch Domestikation verursachte Störungen arteigenen Verhaitens. *Zeitschrift für angewandte Psychologie under Characterkunde* 59:56-75. [Quoted in Cloud, W. 1973. Winners and sinners. *The Sciences* 13:16-21.]
9. Harris, M. 1968. *Culture, man and nature.* New York: Cromwell.
10. Birket-Smith, K. 1959. *The eskimos.* 2nd ed. London: Methuen.
11. Sahlins, H. 1972. *Stone age economics.* Chicago: Aldine-Atherton.
12. Krader, L. 1968. *Formation of the state.* Englewood Cliffs, N.J.: Prentice-Hall.
13. Fried, M. H. 1967. *The evolution of political society.* New York: Random House.

14. Morgan, L. H. 1877. *Ancient society*. Chicago: Kerr.
15. Oakley, A. 1973. *Sex, gender and society*. New York: Harper and Row.
16. Tresemer, O. 1975. Assumptions made about gender roles. In *Another voice,* eds. M. Millman and R. K. Kanter. Garden City, N.Y.: Anchor Press.
17. Fox, R. 1972. Alliance and constraint: Sexual selection in the evolution of human kinship systems. In *Sexual selection and the descent of man 1871-1971*, ed. B. G. Campbell. Chicago: Aldine.
18. Tiger, L., and Fox R. 1971. *The imperial animal.* New York: Holt, Rinehart and Winston.
19. Mead, M. 1935. *Sex and temperament in three primitive societies.* New York: William Morrow and Company.
20. Reiter, R., ed. 1975. *Toward an anthropology of women.* New York: Monthly Review Press.
21. Livingstone, F. 1968. The effects of warfare on the biology of the human species. In *War: The anthropology of armed conflict and aggression,* eds. M. Fried, M. Harris, and R. Murphy. Garden City, N.Y.: Natural History Press.
22. Braidwood, R. J. 1948. *Prehistoric men.* Chicago: Chicago Natural History Museum.
23. Peake, H., and Fleure, H. J. 1927. *Hunters and artists.* Oxford: Clarendon Press.
24. Kety, S. S., Rosenthal, D., Wender P. H., Schulsinger, F., and Jacobsen, B. 1975. Mental illness in the biological and adoptive families of adopted individuals who have been schizophrenic: A preliminary report based upon psychiatric interviews. In *Genetic research in psychiatry,* eds. R. Frieze, D. Rosenthal, and H. Brill. Baltimore: Johns Hopkins University Press.
25. Trivers, R. 1971. The evolution of reciprocal altruism. *Quarterly Review of Biology* 46:35-57.
26. Trivers. R. 1974. Parent-offspring conflict. *American Zoology* 14(1):249-64.
27. Huxley, J. 1932. *Problems of relative growth.* London: Macreagh.
28. Gould, S. 1966. Allometry and size in ontogeny and phylogeny. *Biological Review* 41:587-640.
29. DeVore, I., Goethals, G., and Trivers, R. 1973. *Exploring human nature.* Unit 1. Cambridge, Mass.: Educational Development Corporation.
30. Hamilton, W. D. 1964. The genetic theory of social behavior. Parts I and II. *Journal of Theoretical Biology* 7:1-52.
31. Foulks, E. F. 1975. Schizophrenia held useful for evolution. *The New York Times,* December 9, 1975, p. 22.
32. Lewontin, R. C. 1967. An estimate of the average heterozygosity in man. *American Journal of Human Genetics* 19:681-85.
33. Lewontin, R. C. 1972. The apportionment of human diversity. In *Evolutionary biology,* eds. T. Dobzhansky, M. K. Hecht, and W. C. Steere. vol. 6. New York: Appleton-Century-Crofts.

At the time that this article was written, the Sociobiology Study Group included L. Allen, J. Alper, B. Beckwith, J. Beckwith, S. Chorover, D. Culver, N. Daniels, E. Dorfman, M. Duncan, E. Engelman, R. Fitten, K. Fuda, S. Gould, C. Gross, W. Hill, R. Hubbard, J. Hunt, H. Inouye, T. Judd, M. Kotelchuck, B. Lange, A. Leeds, R. Levins, R. Lewontin, M. Lieber, J. Livingstone, E. Loechler, B. Ludwig, C. Madansky, M. Mersky, L. Miller, R. Morales, S. Motheral, K. Muzal, M. Nestle, N. Ostrom, R. Pyeritz, A. Reingold, M. Rosenthal, D. Rosner, H. Schreier, M. Simon, P. Sternberg, P. Walicke, F. Warshaw, and M. Wilson.

Epilogue

The articles in this volume are critical of a world view that is dominated by "biological determinism," particularly the scientific proclamations that arise from such a world view. Does such a critical attitude lead to rejecting biology altogether? Are we merely cursing the darkness while others are trying to light a match? We think not.

It is often assumed that those who oppose biological determinism must be "environmental determinists," that they believe human behavior and social institutions to be totally a product of the environmental influences to which one is exposed in development. According to this view, generally associated with the psychologist B. F. Skinner, we are simply the result of the conditioning we receive. In place of a human behavior moulded by natural selection, environmental determinists would give us a human behavior moulded by the stimuli and reinforcements of our early lives. To consider this alternative is to pose the question as simply an updated version of the tired old nature/nurture debate—genes versus environment, natural selection versus social science—an institutionalized conflict of academic disciplines, each claiming to explain the essence of humanity. As such, it leads "unbiased observers" to the typical compromise solution: identify the two extremes and assume the truth lies between them. Thus, our behavior is determined partly by genes, partly by environment. The only argument then is how much of each.

To see the error of this line of thinking consider the following example. Imagine yourself living at a time when the process of photosynthesis was not known. Suppose that in attempting to understand the phenomenon of plant growth, a group of scientists did numerous experiments in which plant growth rate was studied under different light regimes. From their experiments, they concluded that plant growth was caused by light. But, at the same time, another group of scientists did numerous experiments in which plant growth rate was studied under different water regimes. They concluded that plant growth was caused by water. A dialogue between these two groups could easily result in an entrenchment along the two different lines of thought rather than a synthesis of the two sets of experimental results. Arguments would be presented on either side to show how the other side was wrong, with scientists being called upon to align themselves with one group or the other.

Finally, an apparent reconciliation proposes that both sides are correct, at least about the experimental results, and that plant growth is not caused by *either* light or water, but rather by both. It would seem reasonable to summarize the debate as not really a debate of substance but rather one resulting from the misguided and narrow interpretation of experimental results. The problem then might be viewed, not as a question of whether plant growth was due to light or water, since it is clearly the result of both, but as what fraction of plant growth is due to light and what fraction is due to water. The debate could then continue, essentially unchanged, with one side arguing that a "majority" of plant growth is due to light and the other side arguing that a majority is due to water, while both agree that the major question is what fraction is due to each factor.

If we knew nothing whatsoever about the process of photosynthesis, the above synthetic argument (what fraction is due to which factor) might seem reasonable. But, with our current knowledge of photosynthesis and plant growth, we see that the so-called synthesis was, in fact, no better than the arguments of the original controversy. To suppose that plant growth could be so "partitioned" (due to either light or water) was fundamentally naive. Although the synthetic argument had a superficial appearance of resolving the conflicts of the controversy, we see from our current perspective that it was just as wrong as the positions of the earlier debate. The entire problem was being approached from a fundamentally incorrect point of view.

The basic structure of this example is equivalent to the biological versus environmental determinist controversy. To see a question of the origin and maintenance of human social behavior or social institutions as merely a choice between determinisms, with the argument being only what percentage is environmental and what percentage is biological, is just as scientifically absurd as attempting to understand plant growth by assigning a certain fraction to light and another fraction to water. Just as water and light are utilized in the process of photosynthesis to drive the complicated biochemical processes that eventually result in plant growth, biological and environmental factors are intimately related in a complex network of interactions between human and environment, which result in the observed patterns of human behavior and social institutions. What then is the alternative to determinism?

The present volume is meant only as a critique of contemporary forms of biological determinism and is, therefore, an inappropriate place for the complete development of a comprehensive theory of human behavior and social institutions. We cannot pretend to present in a two-page summary what centuries of philosophers have attempted to deal with in countless volumes. But we can, at least, outline what we believe to be some of the basic prerequisites for the development of such a theory. It is obvious that the human being and his or her environment are so intimately related that it makes little sense to speak of one without reference to the other. Furthermore, changes continually occur in the relationship of human beings to their environment as a result of their interaction. Thus, we must reject assumptions that relegate human beings to passive recipients of biological *or* environmental factors and develop a theoretical framework that acknowledges human beings as active participants in the moulding of the continual changes that occur in our relationship to one another and our environment. We are not limited by our biology *or* our environment; we actively control the way in which our biology and environment interact. Indeed, it is a reflection on both natural and social science in our culture that they view their contributions to human welfare as telling people what they *cannot* do. Why is it that we do not have a science of human behavior that encourages people to expand their hopes for humanity rather than to contract them? The deterministic alternative is not realism, but cynicism.

A fully developed theory of human behavior and human social institutions will not be achieved by an elitist science that claims to be objective and apolitical. The task of achieving such an alternative theory cannot be separated from the larger task of establishing a just society. That alternative theory will develop along with meaningful social change, the two being inseparable. Indeed, the fullness of human potential will be discovered not by the debates of the academics in our present society but by the conscious efforts of people to build a better society. The words of Emma Goldman ring true still today(1; pp. 61-62):

Poor human nature, what horrible crimes have been committed in thy name! Every fool, from king to policeman, from the flatheaded parson to the visionless dabbler in science, presumes to speak authoritatively of human nature. The greater the mental charlatan, the more definite his insistence on the wickedness and weaknesses of human nature. Yet, how can anyone speak of it today, with every soul in a prison, with every heart fettered, wounded, and maimed?

John Burroughs has stated that experimental study of animals in captivity is absolutely useless. Their character, their habits, their appetites undergo a complete transformation when torn from their soil in field and forest. With human nature caged in a narrow space, whipped daily into submission, how can we speak of its potentialities?

Freedom, expansion, opportunity, and, above all, peace and repose, alone can teach us the real dominant factors of human nature and all its wonderful possibilities.

Reference

1. Goldman, E. 1970. Anarchism—What it really stands for. In *Anarchism and other essays.* New York: Dover.

Suggested Readings

Alland, A. 1972. *The human imperative.* New York: Columbia University Press.

Block, N. J., and Dworkin, G., eds. 1976. *The I.Q. controversy.* New York: Pantheon.

Bookchin, M. 1971. *Post-scarcity anarchism.* San Francisco: Ramparts Press.

Bowles, S., and Gintis, H. 1975. *Schooling in capitalist America.* New York: Basic Books.

Chase, Allan. 1977. *The legacy of Malthus: The social costs of the new scientific racism.* New York: Alfred A. Knopf.

Commoner, B. 1976. *The poverty of power.* New York: Alfred Knopf.

Editors of *Ramparts.* 1970. *Eco-catastrophe.* San Francisco: Canfield Press.

Fromm, E. 1961. *Marx's concept of man.* New York: Frederick Ungar.

Gillie, O. 1976. *Who do you think you are? Man or superman—The genetic controversy.* Saturday Review Press/E. P. Dutton and Company.

Haller, J. S., and Haller, R. M. 1974. *The physician and sexuality in Victorian America.* Urbana, Ill.: University of Illinois Press.

Kamin, L. 1974. *The science and politics of I.Q.* New York: Halsted Press.

Money, J., and Ehrhardt, A. S. 1972. *Man and woman, boy and girl.* Baltimore: Johns Hopkins University Press.

Montagu, A., ed. 1973. *Man and aggression.* 2nd ed. New York: Oxford University Press.

Reiter, R., ed. 1975. *Toward an anthropology of women.* New York: Monthly Review Press.

Sahlins, M. 1976. *The use and abuse of biology.* Ann Arbor, Mich.: University of Michigan Press.